Down a Dirt Road
Reflections of a Farm Wife

Erin Slivka

Down a Dirt Road: Reflections of a Farm Wife
Published by Flax Coulee Press
6074 PN Bridge Rd
Winifred, MT 59489

ISBN 978-0-615-41189-7

Words and Photos Copyright © 2010 Erin Slivka

All rights reserved. No part of this book may be reproduced or transmitted in any form or by any means, electronic or mechanical, including photocopying or recording, without permission in writing from the publisher.

Most articles originally appeared in *The Prairie Star*, a publication of Lee Agri-Media, 1998-2010.

Contents

1. The Cowboy — 1
2. What a Farm Wife Dreads — 5
3. Collision of Two Worlds — 9
4. A Weighty Matter — 13
5. Defining Agriculture — 17
6. Squash Pusher — 21
7. The Object of His Affection — 23
8. From First to Last — 27
9. A Tale of a Tail — 31
10. Laying a Foundation — 35
11. Powdered Sugar Donuts — 37
12. Sold to the Highest Bidder — 41
13. A Swather in the Swamp — 45
14. A New Decade — 49
15. Becoming a 4-H Mom — 51
16. The Passenger — 55
17. Just Call Me Mrs. — 57
18. The Lessons of Motherhood — 61
19. A Disposable Society — 65

20. Amusing our Ancestors	67
21. Barn Cats	71
22. Realistic Resolutions	75
23. Chickens	79
24. The Equipment Itch	83
25. Laundry Lessons	85
26. Looking Good on Paper	89
27. Modified Housekeeping	91
28. The Price of Progress	95
29. A Working Vacation	99
30. The Story of My Life	103
31. Tending Lucille's Garden	107
32. Foodies	109
33. The Importance of Being Specific	113
34. Bad News	117
35. Hometown Heritage	119
36. Scour Guard in My Stocking	123
37. Battling Bovine Racism	127
38. Saving Montana	131
39. The Farmer and the Cowman	135

40. Wedded Bliss	139
41. To Love, Honor, and Cherish	143
42. Farewell to a Faithful Friend	147
43. The Parts Run	151
44. Pomp and Circumstances	155
45. The Magic Kitchen Table	157
46. Children of Summer	161
47. Country Kids in the City	165
48. A Woman's Place is in the Home	169
49. Before You Take Another Bite	173
50. Disappearance	177

The author at the end of the dirt road of her childhood

My Granddad, Art Paulsen

The Cowboy

I was not quite three years old when my granddad suffered fatal heart attack at the age of 64. I don't remember what he looked like. My memories are the disconnected snippets of contact I had with him combined with the deep sadness that washed over my parents following his death.

I knew he was a good man, even at the tender age of almost three. I knew that with a certainty because he held me on his lap and let me eat ribbon candy out of the crystal bowl on the kitchen counter, and only a very, very good man would allow such a privilege.

I don't remember the sound of his voice, but I remember his smile. I can vividly recall the snaps on his shirt and the tangy smell of the tobacco he smoked in his pipe. As the youngest of the grandchildren at that time, I had the distinct place of honor on his lap when we visited, and I always felt right at home.

Tobacco is not the only scent that elicits memories of my granddad. To this day, it only takes one whiff of horse manure to take me back to a fleeting memory of his hands hoisting me up on a horse with no saddle or reins. He taught us all how to guide a horse using its mane and the gentle pressure of our legs against its side, and I remember circling the riding arena with a death grip on the mane of the bareback horse.

In his younger years, Granddad was a cowboy, and in using that term I do not mean he wore a hat and rode horses from time to time. He was the type of cowboy who rarely lived in one location for longer than a year, and he earned a living on his horse. He was nearly 30 when he met my grandma, who was cooking for the cowboys on the ranch he happened to be riding for, and they were soon married. They settled down in a remodeled chicken coop, and he continued to work for ranches here and there to provide for their growing family.

They eventually settled down in Ryegate, Mont., where Granddad trained horses at their place south of town. He gained notoriety for his horse training guarantee: six weeks after leaving a horse with him, a person could mount it from either side, crawl between its legs and sit under it, ride it double or bareback, slide off its hind quarters, and ride it using only a halter.

His work was featured in an article in the *Billings Gazette* and was eventually published in a book by Dan Halligan, who spent a week with my grandparents in 1970. He described Granddad's knowledge of horses as "fantastic." In the *Gazette* article, Halligan said, "If ever a man had 'horse sense,' Paulsen does. He knows exactly what the horse is thinking – sometimes before the animal has the thought."

I find myself longing for some of that ability as I raise my kids; how convenient it would be to know what they are thinking. Alas, that gene seems to have skipped my pool, despite my grandmother's attempts to continue Granddad's legacy and teach us all good horse sense.

I have avoided the wanderlust as well; I'm content to stay where I am for the next 70 years. Granddad was most comfortable on a horse; I am most comfortable in front of a computer.

While I don't have much of a connection with Granddad, his presence has always been apparent in our family. He is here in the thin gold band on my grandma's ring finger and in the photo albums full of horses. His influence is still strong throughout the family. And although I don't have many memories of him, and remain certain that he was a good man. He never had much money or material possessions, but he worked hard enough to move out of the chicken coop and to provide a good life for his family. They sometimes lacked running water, and they all worked hard, but the respect they had for the head of their family was indicative of the quality of his character.

Probably the most telling evidence of who my granddad was is the fact that his home was always open to visitors, and their house was a gathering place for neighbors to play cards, talk horses, and eat Grandma's incomparable cooking.

Although my memories aren't abundant, they are definitely enduring, and he clearly made a mark on my young soul. I have never been ashamed to proclaim that my granddad was a cowboy.

Never trust a man who says he just needs your help for "a minute," especially if he has his spurs on at the time.

What a Farm Wife Dreads

If farm wives from across the world were to gather in a central location and compare their experiences, I am sure they would find that their lives are in many ways the same.

They would probably have similar stories of stocking up on groceries, cleaning grease and manure on jeans, and shopping for clothes at farm supply stores.

They would also agree that there are some phrases that a farmer says to his wife that she dreads hearing.

"Oh, look! Here's an auction sale bill!"

Auction sales are risky due to two factors. First, they place a group of males in a competitive environment, and no one likes to lose. Secondly, they offer a variety of items that sometimes sell for such low prices that a farmer can convince himself that he actually needs them. That's why he will come home with a box of junk that will sit in his own shop until he accumulates so much that he will have to have an auction of his own.

"You got a minute?"

This question is dangerous because by affirming that she has a minute, the farm wife is actually committing to up to three hours of tackling an unpleasant task.

"Did you make something good for lunch?"
In my farmer's language, this question means that he has invited someone else to eat. That someone is standing right next to him, and he is using a code to tell me to put away the peanut butter jar and whip up something else in the five minutes it will take them to arrive at the back door.

"Do you know how much money I just saved us?"
Much like the preferred card savings at the chain grocery store, this question is a ruse to disguise how much money he just spent.

"Are you busy?"
Farmers are blessed with selective vision that allows them to see past the baby on her hip, the boiling pot on the stove, the dish towel in her hand, and the toddler behind her who is emptying the soup cans out of the cupboard.

"Where is last year's calf contract?"
The answer to this question could be "in the calf sales file in the file cabinet," or it could be "I don't know because you took it out last month and stashed it in your calving record book and then took it out to the barn." One thing is certain: last year's calf contract is always in the last place that you look.

"I need your help."
This statement elicits dread because the situation must be bad for him to admit he needs help. The word "need" differentiates this request from the questions like "Are you busy?" and "You got a

minute?" I have found that he only "needs" help when it's a very hard calf pull, a piece of machinery stuck very badly, or a crop insurance policy that needs deciphered.

"Just lift me up in the tractor bucket."
It is unfair to put a farm wife at the controls of a mechanism which could drop her beloved 20 feet to the ground with one wrong pull of the lever.

"How much money is in the checking account?"
Another unfair situation that a farm wife faces is choosing to either lie about the account balance or realize that by telling the truth, she has just given her blessing for him to buy another piece of machinery.

"Help me back this trailer in."
One of the unofficial roles of a farm wife is trailer backer upper. She must position herself in the view of the rearview mirror while the farmer follows her hand signals to precisely place the trailer in the position in which he wants it parked. She ascertains that information after three attempts at parking the trailer which all end with his dissatisfaction. Only when she threatens to begin using a different hand signal altogether will he decide that the trailer is probably okay where it is.

"Don't let her through the gate!"
This urgent command is always yelled across the corral as the farm wife faces down a mad cow with plenty of momentum who is headed in her direction. She can either choose the wrath of the cow or the wrath of the farmer. She usually chooses the wrath of the farmer because he has to forgive her before supper.

To some, it's a pretty picture. To others, it's just another day at work.

Collision of Two Worlds

This morning, two men set off for work.

One rose from bed when the sun gleamed into the window around 5:30 a.m. He donned yesterday's dirt-caked clothes, drank a few swigs of coffee, listened to the weather forecast on the country radio station, and stepped outside, breathing the sweetness of the dewy grass on his way to the corral.

He spoke his first words of the day to the dog, and the next living creature he saw was his horse, which he saddled, mounted, and spent the next few hours upon.

A thousand miles away, the second man awoke to the electronic blare of his alarm clock at 7 a.m. He rose from bed, showered, shaved, and watched the morning news on television. He stepped out the door into the hallway and rode the elevator to the bottom floor, where he departed through the glass doors and onto the sidewalk. A blast of noise greeted him from the dirty street, and he ducked into the corner shop and paid $3.75 for a sugared cup of coffee in a Styrofoam cup.

He then stepped onto the pavement, into a vehicle, and out again when he reached his office building. He took the stairs to his sixth-floor office, where he conversed with at least a dozen people and spent the rest of the morning on the phone and at the computer.

At 4:30 p.m., the office worker will shut down his computer and begin his commute home to his family. He will purchase supper through a drive-through window and return home, having never set foot on the actual earth during his entire day.

The rancher will head home as darkness sets in, sitting down to a dinner of beef that, six months ago, he fattened with grain from his own fields.

The two men live in worlds that would seem to never meet. But now, as travel and communication between worlds apart improve, these two worlds have begun to collide.

The man who spends entire months without stepping on ground that isn't concrete, pavement, or turf will discover that he is missing something in his life. He will begin to search for it, and he will find it in the world of the rancher. He will call it nature or frontier or paradise, and once he has tasted it, he will want it.

He will demand that it belongs to all people, not just those who have worked to preserve it and improve it for centuries. He will demand that it return to its natural state, and he will donate money to organizations that promise to make this happen.

The rancher, who has little money and even less time, will watch defenseless as his world begins to change. When reintroduced wolves begin to feast on his new calf crop, he is told by people living in cities built over the forests where wolves once roamed that it is nature's way. When urban organizations begin to pressure government agencies to pull his grazing lease, he is told that public land is meant for the people's enjoyment, not cattlemen's gain.

He listens in frustration to the television news programs which accuse him of freeloading off the government, while the city dweller pays a smaller percentage of his paycheck for groceries than any other working man in the world. He wonders why these people want to eat foreign-raised meat and grains and conserve for recreation the land that now raises the world's best quality and safest food supply.

He wonders why these people who rarely set foot on the earth itself seem to think they can take better care of it than he can.

And he will continue to wonder tomorrow, as he puts in another 16-hour day of tending his land and his animals. For him, resisting the collision between the two worlds is not just about saving his job; it's about saving his life.

Never underestimate the wrath of a female.

A Weighty Matter

January is not my favorite month.

The beginning of the year coincides with the disappearance of the last of the Christmas cookies, which happen to be my favorite breakfast food. Sometime during that first week of the New Year, I discover that my poor dietary choices have led to the inevitable side effect of shrunken jeans. I can usually hide the resulting roll over my waistband by layering a vest, sweater, or sweatshirt over the top of my shirt, but this year my plan was foiled.

I'm much too frugal to buy a bathroom scale, so the only time I ever have to face reality is when I visit the doctor. Thanks to my medical insurance plan that kicks in only after we have had to sell a couple of vehicles to pay a hospital bill, I seldom visit the doctor, so I have been able to live in ignorant bliss. Until now.

My husband recently purchased a scale for the feed wagon, citing the need to carefully mix the ration for the replacement heifers through the winter. I didn't consider the ramifications of his purchase until he began requesting my help with feeding recently.

Now that three kids are in school full time, that leaves just one little one to occupy the buddy seat in the tractor, leaving me no real excuse to avoid the chores. That's how I came to find myself in the back of the feed wagon, pitchfork in hand, and at the mercy of the shiny new digital scale.

My husband, whose duty is to fill the wagon with the skidsteer loader, began punching buttons on the scale. His eyebrows raised, he glanced at me, and then he began chuckling. I looked at the scale to see what was so funny, and there it was. On the display screen, for all the world to see, was a number I would never have believed to be associated with my weight.

I can understand a little weight inflation based on the fact that it was -5 degrees out there, and I was appropriately bundled up in insulated coveralls, heavy insulated boots, and several sweatshirts under my winter coat. And surely the ear flaps on my wool cap will add some weight. But the number on the display was still significantly more than what I would have expected, and I could feel my face begin to redden.

My husband must have become alarmed as he watched my reaction, because he quickly glanced at the pitchfork in my hand and high-tailed it for the safety of the skidsteer cage. As he dumped loads into the wagon for me to level with the pitchfork, I'm certain I saw him laughing, but he would never come close enough for me to tell for sure.

Later, after the feeders were full and I was pulling the wagon back to the yard, I glanced in the tractor's rearview mirror. My husband was hanging there on the wagon, hitching a ride back. We've had a particularly icy winter, and walking even a short distance is a treacherous endeavor. He looked pretty relaxed back there, despite my lack of expertise in tractor operation, with one arm hanging over the wagon's side while he glanced around, taking in the scenery.

I'm sure it wasn't my fault when the tractor abruptly accelerated at precisely the same time that the wagon was traveling over that long patch of rock hard, frozen cowpies just before the big dip in the pasture.

I'm just not the vindictive type.

Defining Agriculture

These days, it seems that everyone is interested in agriculture. Analysts, economists, journalists, psychologists, and even suburbanites are all chiming in with opinions about how food should be produced. As the interest in food production grows, so does the vocabulary associated with agriculture. As an observer of that trend, I have compiled a helpful list of definitions for the farmer who may be befuddled by the new buzzwords.

- Diversification: sending the wife to town to find a job with a steady paycheck.

- Agritourism: convincing city folks to visit the farm to pick fruit, get lost in the cornfield, or gather the cattle from the north 40, all while paying you for the privilege.

- Value added: the explanation your neighbor gave you for retaining ownership on the steers he didn't get sold last fall.

- Alternative energy: the source of the burst of speed you experience when the cow you have nicknamed "The Old

Bat" comes at you with her nostrils flared when you try to tag her calf.

- Homestead: a couple of acres in New England upon which someone is raising 15 chickens and a goat and calling himself a farmer.

- Conservation: making sure you don't flush or do dishes when the cows are watering.

- Ecofriendly: refraining from cursing at the environmentalists who are explaining why you should embrace the reintroduction of wolves.

- Factory farm: according to anti-agricultural sources, this includes anyone who raises more animals or acres than the guy who is homesteading (see definition above).

- Green:

- Natural: any food not raised on a factory farm.

- Organic: the lettuce in my garden that I never got to eat because the grasshoppers devoured it.

- Corporate farm: see "factory farm."

- Industrialized agriculture: the evil plot by farmers to own enough acres to actually make a profit now and then.

- Foodie: someone who likes to eat, who always buys organic, and who has no idea how to produce enough food to feed the world.

- Green: around here, it's usually the color of the stuff

someone stepped in while during chores and then wiped across the rug in the house.

- Locavore: someone who lives in an area populated enough to support a farmer's market where local foods are available.

- Climate change: what every Montana farmer prays for in January.

- Sustainable agriculture: the hope I have that the farm will sustain our family despite the threats of those whose definition of "sustainable agriculture" differs from my own. Their philosophy is that a farmer should produce enough food to feed the world while not using chemicals that might harm the environment, not offending the neighbors with smells and noise, not taking government subsidies, and not having the audacity to hope for a profit. And unless that philosophy is altered, American agriculture may not be sustainable at all.

Squash Pusher

After years of struggling with drought, heat, bad water, grasshoppers, deer, and toddlers, I vowed to never again attempt gardening. I was resigned to a summer of purchasing vegetables at the grocery store and looking forward to a reprieve from my war with kochia.

When spring rolled around, I yielded to the pleas of my children and helped them plant a few packets of seeds. While I was transplanting three tomato plants, the kids were happily burying seeds at the other end of the garden. Only when they reached the end of the row did I realize with horror that I had turned them loose with the zucchini seeds, and they had none left.

Knowing that a single plant could produce an annoying amount of squash, I had frightening visions of myself becoming a zucchini pusher on the streets of our small town, forcing squash upon my unsuspecting neighbors and lurking in the church parking lot where no one locks their car doors.

I thought I might be saved from such a fate when a late frost struck down several plants, but the zucchini persevered.

The cucumbers wilted and died, and the tomatoes fell victim to the deer. Sunflowers cropped up in various places, undoubtedly the products of a small person distracted by the cats. Kochia filled in the spaces between the rambling pumpkin vines.

I watched the growing number of zucchini plants with concern, but when I quit watering them for fear of spoiling the soil with our salt laden well water, I figured I had little to fear.

Only when we returned home from a short vacation did I remember the real reason that I had ceased to garden: you reap what you sow.

Keeping up with the produce from my tiny garden plot has become quite a challenge, and it turns out that my nightmare of becoming a zucchini pusher has become a reality. With eight to 10 of the squash on my counter at all times, I have been reduced to begging my family and friends to take them off my hands.

I bought a food processor so I could grind the stuff up small enough to hide it in casseroles, brownies, cookies, and cakes. I have scoured the Internet for recipes and have stuffed zucchini bread down my children for three meals a day. I have fried it in so much oil that all the nutritional value has been sucked out and replaced by cholesterol.

One would think that I would just give up and throw it out, but I have enough of my parents' thriftiness instilled in me that I cannot bear to do it, so I press onward. I stopped short, however, of making the recipe I discovered for zucchini gravy.

It seems that zucchini and wheat have a lot in common this summer; they are plentiful, but the market just isn't there.

Next year, I may re-evaluate my summer pastime. Considering the looks on the kids' faces when they recently saw the size of their pumpkins, though, I will probably be in the same predicament come next August.

Anyone want some zucchini?

The Object of His Affection

One of the benefits of living in the middle of nowhere is that you seldom have to worry about competing for the affection of your spouse.

You don't really have to worry about your husband running off with the neighbor's wife when she lives two miles away and happens to be his mother.

However, it didn't take me long to figure out that other women weren't going to be my primary source of competition for my spouse's attention. Soon after we were married, I realized that my competition had dark brown eyes, shiny hair, a gentle disposition, and a desire to always please the master of the house.

There was simply no way that I could compete with the object of my husband's affection, even if it was a Border collie mutt.

The whole affair started out innocently enough. The two of them would spend most of the day together, and the hours just seemed to get longer and longer. They would come into the house and share their supper together, talking about the day they had spent side by side.

I began to envy my husband's canine companion. It seemed that he was receiving much more attention than I was, and I began to see other warning signs as well.

When I walked to the pickup to accompany my spouse, the dog leapt out in front of me and secured the seat next to my husband for himself. I spent the trip wedged against the passenger door while the dog snuggled up to my husband and licked his face. Every now and then, he would turn toward me with a smug look upon his smiling face and wag his tail, causing clouds of dust to rise up into my face and make me sneeze.

There were other signs of trouble as well. Occasionally, the dog had to stay at the house while my husband worked in the field. When the master returned at the end of the day, both the dog and I would greet him at the door. At first, I received the warmest greeting, but the canine companion soon began to receive the most enthusiastic hello. The two of them would smile, embrace, and engage in a round of rough-and-tumble play while I stood at the door watching the display of affection.

They soon began to flaunt their relationship in front of me. At the dinner table, the dog would sit next to my husband and was often treated to the prime cuts of meat that I had prepared for my spouse. Once the dog's access to the house was limited to the porch, but he had soon weaseled his way into the living room. Eventually, he even began spending an occasional night at the side of his master next to the bed.

I have tried to rationalize my jealousy. What threat is really posed by a 40-pound mutt who eats rotten roadkill and rolls in manure? What kind of intellectual stimulation can he really provide when the highlight of his day is sneaking scraps out of the cat dish?

This line of thought began to bother me, though. A critter with horrendous breath and half the dirt in Fergus County lurking within his fur was actually winning the competition for my spouse's affection.

Perhaps I could pick up a few pointers from the dog. I guess everyone could be reminded to always greet your loved ones with enthusiasm and don't withhold your feelings of affection.

I think I'll stop short of imitating his personal hygiene habits, though.

The Slivka kids pose on the last first day of school as Emma Lou starts kindergarten. Clockwise from upper left are Anna, Riley, Matthew, and Emma Lou.

From First to Last

At some point in the last year, I crossed a line in parenthood.

I can't define the moment at which it happened. I can't even recall the week. It may have been a slight shift over the course of a few months.

All I can say for sure is that a year or two ago, I was focused on firsts. I was anxious for the first smiles, first foods, first words, and first steps. I was always impatient for what was just around the corner. I told myself that life would be so much easier when the naps were more predictable, when the nursing was finished, when they were all dressing themselves, and when they were all in school.

I thought it would be nice to wear a shirt without spitup on it. I thought it would be lovely to shower without little fists pounding on the bathroom door. I was certain that breakfast would be more pleasant once I was no longer required to spoon the cereal into someone's mouth and wipe it off my face when someone blew raspberries with a mouth full of oatmeal.

I remember my excitement at the prospect of kids who could buckle themselves, wipe themselves, read to themselves, and wash themselves.

More experienced moms would tell me to slow down and savor the moments, but I dismissed them with thoughts that surely they must be going senile.

How could anyone not look forward to the time when the children could independently tie their shoes? Who doesn't look forward to the completion of potty training? Do these women actually miss the endless nights of laundry during the bouts of stomach flu or the hours of lost sleep because the baby won't stop crying?

And then I realized that, instead of experiencing so many firsts with my kids, I was beginning to experience lasts.

When was the last time I read *Goodnight Moon* three times in a row? When did I last receive an enthusiastic "glad to see you" hug from my oldest son? When was the last time I washed my daughter's face and smooched her nose?

Granted, there are a few lasts that I will not mind marking: the last diaper, the last pet beetle in a jar, the last episode of *Dora the Explorer*, and the last spilled milk that drips into every crevice of the table.

But when was the last time I held a newborn who scrunched up under my chin and became an extension of me? When was the last time I watched someone smile in her sleep? When did I last rock a little one to sleep and feel the heavy weight of my parental responsibility on my chest as he rested, completely secure in my arms and dependent upon me for his survival?

At some point, there will be more lasts, like the last request to sleep with Mom, the last first day of school, and the last t-ball game. We will experience the last bedtime story, the last sleepover, and the last game of hide and seek. Some of those lasts will pass by unnoticed. Some will bring a tear.

I know that, as I pass from a mom focused on firsts to a mom considering lasts, I will try to savor the sweetness of this stage in life. Instead of wishing for the next first, I'll spend some time enjoying the here and now. I'll probably still cry at graduation this year, knowing that someday it will be me saying goodbye to my babies, but I'll attempt to make the most of the fleeting time I have to create memories with them – both firsts and lasts.

A Tale of a Tail

$\mathcal{I}$ began my day with a nice shower. This is not a practice that I take for granted. Many times my personal hygiene plans are derailed by a baby who wakes up too soon, a phone call that changes my schedule, or a lunch that needs packed. And with small children in the house, it's always risky to step into the bathroom for 15 minutes and leave them, in all their imaginative and exploratory glory, unattended.

That risk is doubled lately because the two-year-old is potty training, and the cost of taking a shower may be a 15-minute cleanup and a load of laundry.

Happily, we were all clean and dry when I emerged from the bathroom.

That changed about an hour later.

Having finished seeding the barley, my husband has now switched back to ranching mode, and he discovered a sick calf this morning. He enlisted my help to doctor the calf. He was not specific about my duties until well into the operation, and wisely so, because I would have refused to leave the house had I known my fate.

The calf in question was bloated. I took one look at his soft, tufted ears and felt a wave of motherly compassion. I was willing to do whatever was required of me in order to make the cute creature well again.

My husband began to talk to me in rancher code. I asked him what the plan was, and he said, "I'm either going to rope it or chase it into that deal over there, but it's not ready yet, so don't let it go past the gray gate."

Right.

Given the fact that there were three gray gates in the direction he was pointing, I was a little confused. I meandered over in that direction and stared blankly for awhile, and then I came back out.

"Are you going to help me catch this calf, or are you just going to stand there?"

Ah, of course. We're catching the calf.

I missed that memo.

So we chased the calf around in circles for a bit until he climbed on top of the manure pile. My husband roped him. The calf took off like a bullet. Somehow the calf ended up close to a post, and Shane wrapped the rope around a post. I helpfully stood a good distance away from the calf so as to avoid contact with his rear hooves, which were flailing like mad.

I suggested another rope to restrain those flailing legs, but my husband poo-pooed that idea. (You'll understand that bad pun very soon.)

He grabbed the calf, flipped him over on his side, and grabbed a rear leg to restrain him. Then he looked expectantly at me.

I looked back and shook my head. There was no way I was going to shove the tube down the calf's throat. With my luck, I'd hit the wrong hole and fill his lungs full of mineral oil. Nope. Not going to happen.

But that meant I had only one other option, and it had to do with the back end of that calf.

I suddenly found my freshly-showered self sitting on the manure-laden ground hanging on to a calf's back leg with all my might. And, as the medication was injected and the tube inserted, I found myself a bit too close to the business end of that calf, which was now spraying excrement in my general direction and flicking it into my freshly-washed face and hair with his tail.

When the doctoring was almost over, my husband decided that he should go get a longer tube to insert into the calf's stomach and relieve the gas pressure. He told me to go ahead and wait there while he went to the shop to retrieve it. I glared at him through my manure mascara.

He returned and tubed the calf again. I was still hanging on. And glaring.

At that point, he decided that the calf may have an infection. He figured he might have to go get him another type of medication. He told me to be sure not to let that calf go while he went back to the shop. Again.

At this point, I noticed the two-year-old was in the adjacent corral. She wasn't wearing shoes. She tried to climb the fence to get to me, but I used my best "Mama means business" voice to convince her that she needed to stay away.

She busied herself playing in the feed trough.

After what seemed like years, Shane returned with the new medication, administered it, and told me to let the calf go.

I was happy to see his behind getting farther and farther away from me. He didn't look nearly so cute anymore. The calf, that is.

I went to check up on the toddler at the feed trough. I was expecting to find some leftover grain in the trough since she was so interested in its contents.

Instead, I found water, a block of wood, and a chunk of dehydrated manure.

As I picked her up, she exclaimed, "Look, Mama! I put dat tow poop in dere!"

You could say that my morning was really crappy.

Laying a Foundation

In 1914, my great-great-grandparents brought their family to Montana, settling down in a log cabin and laying the foundation for a ranch and a way of life that would stand the test of time.

I'm not sure I understood the significance of their imprint on my life until recently. As a child, I was most aware of the influence of my parents. Unlike most of the parents of my friends, my folks were quite unreasonable.

I was forced to share not only a room but a bed with my older sister, who would draw an imaginary line down the mattress and force me to huddle on my 1/8 of the bed with a sliver of blanket.

They made me play for hours outside with little entertainment aside from siblings, acres and imagination. They shushed me when the weather came on KGHL every morning at breakfast. They took us shopping at the thrift store and Big R while my friends visited the mall.

My parents threatened to punish us if we disobeyed, and then they followed through. They made me clean my plate, respect my elders, do my chores, use my manners, and take responsibility for my wrongdoings.

They had high expectations and low tolerance for excuses.

Not until many years later did it occur to me that my parents didn't just happen to adopt these philosophies of child rearing. My parents and grandparents and great-grandparents had simply built upon the foundation begun in 1914.

As I retraced the steps of my grandpa's childhood in my backyard, some of the elements of his upbringing had seeped into my own. Even if we cannot comprehend the harsh realities of farming and ranching in a time of hand pumped water, no electricity, and reliance on horses and wagons, we can certainly appreciate the sacrifices of the previous generations and strive to pass that appreciation on to the next generation.

In some ways, I envy those earlier generations of parents. They were not competing with laptops and cell phones for their children's attention. The television was not the focal point of family time in the evening. Kids who worked hard during the day were grateful for whatever meal was offered them in the evening.

But even as I wonder what my kids could possibly have in common with the children of my grandpa's generation, I remember those philosophies that I can see reverberating throughout my own parenting, and I recognize the imprint of that family foundation established so long ago.

The foundation, I am certain, is a solid one. In a few days, we will gather in the mountains above the place my ancestors began their new life in 1914, the same place where I spent my childhood. My grandpa will be there, as he has been every year on July 4 since he was a boy. We will have fried chicken, watermelon, and homemade ice cream, just as the family had 80 years ago.

We will celebrate freedom, tradition, family, and heritage, and I'll make sure I find the time to reflect on those who took the risks and worked the land so that we might enjoy the rewards of their foundation.

🌾 Powdered Sugar Donuts

My grandma let me eat powdered sugar donuts, and she didn't even flinch when the white powder misted over the green vinyl tablecloth and chunks of cake tumbled to the floor.

At her kitchen table, I tasted my first bowl of Froot Loops. My mom bought only Rice Krispies and Cheerios. Being allowed to eat brightly colored, sugar laden cereal was, in my five-year-old brain, a sure sign of my grandma's affection for me. And when she pulled the package of Oreos out of the cupboard, I was certain that I was adored.

Grandma bought me ruffled dresses for my birthday, read *The Poky Little Puppy* Golden Book at least a thousand times, and rocked me to sleep while singing "Bimbo." I cannot recall a time when Grandma lost her patience with me, not even when she rocked me through the night when I was feverish and miserably suffering through the chicken pox.

She sang when she washed the dishes and told me stories of learning three-part harmony while standing at the kitchen sink with her sisters. When the dishes were done, she let me plunk at the piano keys and then laughed when I spun myself dizzy on the piano stool. Then she would sit at the piano herself, opening up her music and playing with the same smooth grace that I saw in her when she waltzed in Grandpa's arms.

She took me to the library and let me spend as much time as I wanted selecting books. When I was finished, we stopped at the grocery store, where she told me to choose any piece of candy that I wanted. The choice was agonizing because it was such a rare treat, and I wanted to both savor the freedom to choose and make the perfect selection.

In my childhood, the things Grandma did for me and allowed me to do were at the forefront of my mind. As I grew up, I began to realize that her impact on my life was far more profound than providing me with powdered sugar donuts.

Grandma's presence was reliable throughout my life. She was there at basketball games, music concerts, and awards ceremonies. At every birthday party and every Christmas Eve, Grandma's presence was a certainty. She drove 45 miles one way to take me to piano lessons, and during those drives we spoke about my future.

Actually, it was a bit of a one-sided conversation. Grandma was a champion of education, and she was determined that her grandchildren would go to college. When I announced my engagement, her statement was simple. "Congratulations. You better be planning to stay and finish college." Her pride was evident when she watched me grasp my diploma.

To read about my experiences with her, a person would think that I was an only grandchild. The truth is that she extended this dedication nine times over, and she had a special connection with her 15 great grandchildren as well.

The evolution of my admiration for Grandma never stopped. My five-year-old appreciation for the sweetened cereal gave way to an appreciation of her beauty and grace. I admired her wit and her ability to retain everything she read or heard, and I was certain that she could defeat any of the *Jeopardy* contestants that she faithfully watched five days a week. But later in my life, when she watched each of my four children at the baptismal font, I realized that she has been my steadfast example of how to live as a faithful follower of Jesus Christ.

Grandma's faith was an inseparable component of her life. She didn't lecture me about how to live as a Christian; she showed me. She didn't tell me how to nurture my children; she had already demonstrated it in my early years. She did not offer marital advice; she griped about Grandpa's habits while dedicating nearly 65 years of her life as his wife. She was not afraid of death. She was thankful for her blessings, had lived a full life, and was ready to meet her Savior.

I only wish that I could let her go with the same grace that she possessed.

My grandparents, Jack and Betty Robertson

Riley has his first taste of tractor fever at age 4.

Sold to the Highest Bidder

Of the many qualities that set country folks apart from city dwellers, their behavior at auction sales must be among the most distinguishable.

While I have never actually been to a big-city auction, I have seen them on television and have read descriptive accounts of the events. Apparently they often include wine sipping and cheese nibbling prior to the main event. Then the bidders quietly assemble in pretty rows of seats and grasp delicate paddles adorned with perfectly printed numerals. When they wish to place a bid, they quietly raise their paddle until they are acknowledged by the jacket-and-tie-sporting auctioneer.

Country auctions are another matter altogether, with the first noticeable difference being that they are held outside where bidders are treated to snow, rain, scorching heat, or gales of wind, depending on the season. Weather rarely daunts the crowd of friends, family, neighbors, and those other folks drawn to the event by a sale bill advertising a particular piece of equipment.

About the only difference the weather makes is in wardrobe and beverages. In the heat, the crowd is a sea of straw hats holding lemonade or beer. In cold conditions, wool ear flapper caps bob throughout the assembly of people holding steaming coffee in gloved hands.

The method of bidding is much different in a country auction as well. Most seasoned auction goers tuck their number, written with a Sharpie on a piece of tagboard, into the unsnapped pocket of their western shirt. When it comes time to bid, only novices wave their number up high in the air. The veterans of the country auction each have their own way of bidding, ranging from the raising of a finger off the Styrofoam coffee cup in their hand to the barely discernable nod of the head.

Auctioneers in the country have, in my opinion, a much more difficult job than city auctioneers. While the country auctioneer might have a less expensive wardrobe to worry about, he is tasked with trying to remember nearly every rancher's name in a multiple-county area since he will usually announce the winning bidder by name, not number.

He must know the value of every piece of farm equipment manufactured in the last 100 years so as to start the bidding at an appropriate value. He must also know the current state of the livestock and grain markets. If farmers are feeling pinched, he goes up in $10 increments. If they are on the verge of selling $1.30 calves, his increments may stretch out a bit.

Almost as crucial as the auctioneer are those who are assisting in the auction. Among them must be someone mechanically talented enough to coax an ancient tractor to start and run long enough to get a bid. In addition, those capable assistants must help the auctioneer group and order items in such a way that people come early and stay late.

Perhaps most distressing to the farm wife is the way in which these folks put a perfectly functional grease gun in a box of assorted gleanings from the shop drawers, most of which are useless. The thrifty farmer cannot resist the idea of a $5 grease gun, so the box of goodies goes home to join the assorted junk in the drawers of its new home. I am convinced that some of these boxes of goodies just sit in a corner until they are auctioned off again in 20 years.

I am also convinced that some of the buyers at the auction sales are bidding more on principle than on need. It is as if they are actually bidding farewell to a neighbor and colleague by supporting their auction.

There is a certain sadness that permeates an auction sale. It is the definitive end of someone's life in agriculture, and many times, it is one more tally mark that indicates the dwindling number of families in an area. The land and equipment is swallowed up by neighbors who must expand to continue to make a living.

Despite the sometimes distressing reasons for the auction sale, there is always the sense of community, and it remains one of the only events that might draw as big a crowd as the weekend football game. That sense of community is perhaps the very reason that we don't live in the city and hold our auctions indoors like the city folks do.

A Swather in the Swamp

As she rounded the corner of their driveway and headed toward home, the farmer's wife noticed a peculiar sight off to the right. There in the wheat field, right next to the duck pond which sprang up to drown the furrows during the deluge of rain this spring, sat the swather.

The presence of a piece of equipment in the field didn't normally concern her, but a swather in the middle of the swamp with its header pointed toward the winter wheat caught her attention. Much like those who have near-death experiences, she was suddenly caught in a vision, although her vision was of the future, not the past. She visualized her afternoon plans evaporating, replaced by the chore of pulling a swather out of a swamp.

The vision was interrupted by the sight of her husband standing along the road with his thumb in the air. After careful consideration of her options, she stopped the vehicle and let him in.

Because she has been married for a dozen years, she knew that she must immediately take responsibility for the problem.

"It's my fault, I'm sure," she declared.

"Yes, it is. You're the one who was complaining about the cheatgrass next to the road. I was only trying to make you happy by cutting it down," came his reply.

It occurred to her that eight or 10 years ago, she wouldn't have been so wise. She would have asked him what in the world he was thinking by driving the swather into the duck pond. She might even have complained about her afternoon plans being wrecked by the task of pulling the swather out. But now, she is older, wiser, and a bit more sarcastic. That's why her next statement was obvious to her.

"I would just like to acknowledge that anything that might happen later on today will also be my fault," she told him. "And I'm sure that something will break."

He nodded his agreement.

"As long as we're clear on that," he said.

She then thought of the young mom who approached her at a recent baby shower to vent her frustrations about all the things she didn't know when she married a rancher.

Few farm and ranch wives will admonish a young bride about the pitfalls of partnering with an agriculturally inclined man. No one wants to spoil her dream of riding off into the sunset on horseback by explaining that she will be left behind to close the gate and pick up the 14 calves that are trying to turn back away from the cows.

No one wants to blemish the vision of romantic candlelight dinners by telling her that if you're eating by candlelight, it's because the power has been off for a few hours and you're dining on a can of beans. When you do have power and the inclination to cook a nice meal, your spouse is likely to be several hours late, leaving you to vacilatte between thoughts like, "I sure hope something hasn't happened to him," and "If he's not laying dead somewhere out there, I'm going to kill him when he gets home."

It just wouldn't be right to tell a blushing bride that he will always come to the door and ask her to help move a piece of machinery to another field just as she is laying the baby down for a nap and looking forward to some peace and quiet. It wouldn't be right to tell her that while her city friends have a week's vacation to Florida in the summer, the only traveling she's likely to do in August is a flying trip to town to replace a broken belt on the combine.

Likewise, it would not be tactful to mention to the young groom that his new wife is likely to make more mistakes than he ever thought possible during her first five years on the place. She will drive off in the truck with the emergency brake smoking all the way down the road. She will ruin a rim by driving a mile on a flat. She will burn the bread for harvest dinner, and she will overfill the grain bin.

It would not be wise to tell a new groom that his wife will have significant trouble learning to decipher his sign language while they are moving cows. When he clearly means to open the gate over by the old Jones place, she will read the hand signals as a command to ride back to the house and pour herself a nice cool drink of lemonade.

When the two are working cows in the corral, she will let all the wrong pairs through the sorting gate. She will constantly be in the wrong place at the wrong time, and he will feel compelled to teach her a thing or two about working cattle. He will be mystified when she reacts to his teachings with either a volcanic fury or a torrent of tears, and he will be equally confused by her indignation that he expects her to have a meal on the table the moment they return to the house.

The wise woman realizes that working cows is a chore that

must be done, and chances are high that she will get yelled at and will have to yell back. Then they will never speak of the incident again.

The seasoned farm wife knows that passing along all this information would be too overwhelming for a new bride. Instead, she should know just three things.

First of all, in the first few months of marriage, a new bride should never do something that she doesn't want to do for the rest of her life. For example, if she admits that she knows how to run a swather, she can expect that she will be responsible for that task for perpetuity. Sometimes it's best not to be too eager to help out.

Secondly, she needs to know that she should have no expectations and make no plans. That way she will avoid the disappointment of missed appointments, weddings, dinners, trips, and checks.

Lastly, as illustrated by the incident of the swather in the swamp, she must always admit fault right away. If she informs her partner that she is guaranteed to miss a gear, turn the wrong way, or stall it out, he is rendered incapable of being angry at her. After all, he was told what to expect. And so, when the tow chain breaks and the farm wife who is in the swather is turning the wheels in the wrong direction, they just might be able to share a laugh instead of an angry exchange.

Disclaimer: The above column is for entertainment purposes only. Any resemblance to an actual person or event is purely coincidental, especially if that person is married to the author.

A New Decade

The last month of the year finds me scrambling to finish projects, finalize farm bookkeeping, and find the perfect Christmas gift for everyone on my list. It's a frenzied time of baking, cleaning, and making sure the girls don't spill anything on their Christmas dresses. I must carefully wrap each gift so I can later pick up the shreds of paper strewn across the living room floor. I painstakingly decorate the tree that usually falls down at least once before Christmas, leaving a pile of broken ornaments and needles below.

I'm ushering kids to Christmas plays and pageants, taking pictures and mouthing their lines to them from the audience. I'm baking treats to deliver to school parties, exercising patience as the kids decorate sugar cookies, making crafts and singing carols.

The Christmas card photo has been taken and the letter has been sent. Somewhere in the midst of the holiday season, another year slips away.

This year is especially difficult to bid farewell. Not only are we entering a new year, but we are welcoming a new decade.I rather liked the old decade. The old decade saw three new babies born. It witnessed the first days of school, the first loose teeth, the first successful hunting trip, and a multitude of hugs from little people in footed pajamas.

This new decade will bring strange new adventures. In the next decade, I will be parenting teenagers. I'll become the mother of an adult. There will be cars and dates and acne and goodbyes.

The next decade will bring 40. It will mark 20 years of marriage. This will be the decade in which I can embarrass my children simply by showing up at their school and acknowledging that I know them.

As many transitions as the next decade holds for my family, it also holds changes for agriculture. In the past decade, we saw mounting challenges to our industry. Increased activism in the areas of animal rights and environmentalism, combined with media bias and social networks, have done great harm to the image of agriculture as well as setting up legal battles. Decision making is becoming more and more difficult due to changes in farm policy and market influences that stretch far beyond the fundamentals of supply and demand.

As producers, our response to these challenges will be the difference between success and failure. Even as we tackle the issues of raising teenagers in the next decade, our most difficult struggles may involve finding a way to continue farming in a world that will demand more and better quality food that will be grown on a shrinking number of expensive acres.

While I'm curious to see what will happen, I'm glad I don't have a crystal ball. I have a feeling that we don't want to see all the challenges that agriculture will face in the next 10 years. But I also have faith that if anyone can meet those challenges, it's our friends and neighbors who, from one coast to the other, work every day to make a living and provide safe and wholesome food. If only I had that much confidence in my ability to raise teenagers.

Becoming a 4-H Mom

In my mind, no one epitomizes motherhood more than a 4-H mom. If you ever want to see a mother's love and dedication to her children in action, visit a county fair about a half an hour before the beef showmanship contest.

Along with my admiration for the 4-H moms is a fear of becoming one. I have successfully avoided the topic thus far by changing the subject whenever someone asks me when Riley, our 10-year-old, will be joining 4-H. I can dodge the topic with such skill that the inquiring person does not even realize what has happened.

But when Riley expressed interest in joining, I was unable to dodge. I was rendered excuseless. And I was struck with the cold fear that I would actually have to *become* a 4-H mom.

Apparently there are many 4-H moms. My sources tell me that there are 6.8 million 4-H members nationwide. Assuming that many of those 6.8 million children have mothers, there are more than a few women who have embraced the idea of 4-H motherhood more readily than I have. In fact, I know a few 4-H moms. They are the women who look very frazzled during the third week in July and are barking directives such as, "Johnny, you have to get out of the pool NOW! Your 4-H interview is tomorrow, and you haven't even started your record book yet!"

My reluctance is not a reflection of the 4-H program itself, which I know to be very beneficial to its members. When all those little voices recite, "I pledge: my Head to clearer thinking, my Heart to greater loyalty, my Hands to larger service, and my Health to better living, for my club, my community, my country and my world," it cannot possibly be a bad thing.

I am a proponent of encouraging kids to tackle projects and responsibilities. I think seeing a project through from beginning to end is a valuable life lesson that the 4-H program successfully fosters in its members. The adult-youth partnerships and the community service components of the program are also benefits that my children could stand to reap.

In fact, I'm a big fan of 4-H. I think it's a great idea that my kids become members.

I'll sign them up just as soon as I find a good 4-H mom for them.

They're going to need someone who can brightly encourage them to work on their record books as they go without nagging them and turning it into an episode of tearful tantrums. They will need someone adept at baking cookies for meetings, entertaining toddlers during said meetings, and "helping" to make those demonstration speech posters at the last minute.

The right person should be able to know how to trim a chicken, train a cat, give proper feed rations to a steer, and show a pig. She should know the secret to washing an animal and keeping it clean while it occupies a 10-square-foot pen in which it eats, drinks, and has nowhere to lay down except in the spot where it just dropped the results of the eating and drinking.

The kids' new 4-H mom should be able to load and unload

a calf and drive a pickup and stock trailer in town. She should also be able to back up the camper to the proper position and cheerfully pack enough clothes and food in said camper so that the family can live there happily for several days. She should enjoy using public showers and washing the 4-H child's only white shirt in the sink.

The ideal mom should be creative and artistic, able to whip up clever display signs and slogans. She must also have a comforting shoulder for the child who is learning about a world in which there are winners and losers.

She should also be familiar with all 110 different 4-H program areas and know the nuances of each well enough that she can coach the child who chooses as his projects **Advanced Visual Arts: Draw/Fiber/Sculpt, Sportfishing III, and Meat Goat I.**

Since I am not likely to have many applicants for this unpaid and underappreciated position, I will likely have to tackle it myself. I will have to face my fears and recognize my weaknesses. I will have to ask questions, learn the rules, and give it my best shot. And in doing that, I will likely be teaching my kids the best lessons of all.

The Passenger

The loss of one person from a small rural community can leave a hole so gaping that every person for miles around feels the wound. So it was in our little town last week.

Tommy was many things to many people, and the magnitude of his life could be seen through the extensive list of his involvements and accomplishments. But the true goodness of Tommy was shown to me through the heart of a little boy.

In his 40 year career of driving a school bus, Tommy had driven hundreds of kids to countless destinations. That makes it all the more remarkable that he would take the time to befriend the little boy who sits all alone at the front of the bus. In the 30 minutes that the two of them spent together as that bus bumped its way along the gravel road to and from school every weekday, Tommy must have answered thousands of questions. The purpose of every button, latch, and knob in the bus was undoubtedly explained to satisfy the curiosity of a little mind that never stops churning. Reports of these conversations would arise at home.

"Mom, Tommy told me how the clutch in the bus works."

"Mom, Tommy says that our old bus is better than that highway bus. And he doesn't like those automatic transmissions."

"Mom, can we listen to 92.5 on the radio? Tommy says that's the best station, and it has the best farm news."

Not only did Tommy unfailingly answer every inquiry, but he made sure that this little mind, so full of the details of how things work, did not forget to arrive home with his jacket, gloves, hat, and backpack each day.

On Wednesdays, the two of them developed their own sign language denoting whether it would be Bible school or a bus ride home at 3:30.

And when it was time for the little boy to descend the stairs of the bus and go home, Tommy would never let him go by without giving him "five." Occasionally, the straight-faced bus driver would offer the first grader a beer and a chew of tobacco, too, which would elicit a crooked grin, the joy of having a buddy to joke with reflecting in his eyes.

So when it was time for him to descend the stairs and walk past the driver's seat occupied by someone else, the little boy paused, as if Tommy could somehow reach out to slap his hand one last time. He stared out the window into the distance for a moment, comprehending the loss that transcended the silence, and slowly stepped off the bus, disoriented and heartbroken.

In his very special way, Tommy made a difference in this young life. He could have chosen to simply drive the bus, but instead, he chose to devote himself to his passengers. He was far more than a bus driver. In a world that can be confusing and overwhelming for my little boy, Tommy was an ally, a caretaker, and a rare friend.

With any kind of luck, that little boy who sits all alone in the front of the bus might someday grow up with just a bit of the goodness of Tommy tucked away inside him. For that, I am so grateful.

🌱 Just Call Me Mrs.

Regular readers of my columns know that I consider myself a common sense advocate. Since common sense is a truly endangered concept these days, I frequently find myself pondering situations in which it is lacking.

In many cases, common sense finds itself lost in an idea that started out with good intention, such as the feminist movement, and then became fogged over and lost all sensibility whatsoever.

The feminist movement presumably started when women realized that they did not have the same rights as men in society, so they began to demand more rights. Logical enough. The problem began when people started using terminology such as "equal" and insisting that no differences exist between the two genders.

Obviously, differences do exist between men and women, and we have very different strengths and weaknesses. That means we excel at different tasks, and it also means (gasp!) that we are not equal. We were designed to complement each other. That's why I cook and my husband changes the spark plugs.

The way I see it, women really messed up when they insisted that they could do anything a man does, and maybe even do it better. Now we're stuck doing all those dirty jobs that we didn't have to do before feminism, and in most cases we still have to do the traditionally female tasks as well.

And, since we are equal now, we decided we could no longer take our husband's name when we married. Instead, we hyphenate our name to make it equal.

Aside from the fact that a single last name provides for family unity and identity, I have another problem with the hyphenation issue. If I, for example, had become Erin Robertson-Slivka when I was married, and if I insisted the kids become little Robertson-Slivkas, what happens when they get married and have children? Riley Robertson-Slivka could marry Jane Smith-Jones, who would then become Jane Smith-Jones-Robertson-Slivka, and by the time my great-grandchildren were born, they could be named John Gibson-Evans-Michaels-Peabody-Smith-Jones-Robertson-Slivka just so we wouldn't slight a female ancestor somewhere along the family tree.

My humble opinion is that this just doesn't make sense.

Neither does the idea of abandoning the prefix "Mrs." so as not to offend a feminist. Honestly, I am offended when I receive mail addressed to the generic "Ms."

I've earned my Mrs., thank you very much.

I have put in 15 good years of marriage. Now, 15 years might not seem like long to some folks, but I feel like it's enough hard work to earn me the title of "Mrs." I have survived many conversations in the tractor with my head clanging against the glass as I sat on the armrest for four hours. I have produced around 6,000 dinners, at least that many sandwiches stuffed in a lunchbox, and hundreds of pans of eggs made with a baby on my hip.

I've endured 39 months of pregnancy, 48 cumulative hours of

childbirth, and seven trips to the ER with ill children.

Our marriage has survived many episodes of working cows together, taking vacations together, and even camping as a family.

We have emerged from countless tax seasons, calving seasons, and hunting seasons with our vows intact.

Yes, I have earned the title of Mrs., as I remind myself when I'm bouncing along on the back of the four-wheeler at eight months pregnant or riding around the yard on the top of the air drill to make sure the power lines aren't going to catch on the wings.

It's just too bad that society doesn't value my Mrs. title nearly as much as it values my B.S. degree, because we all know how much that is worth.

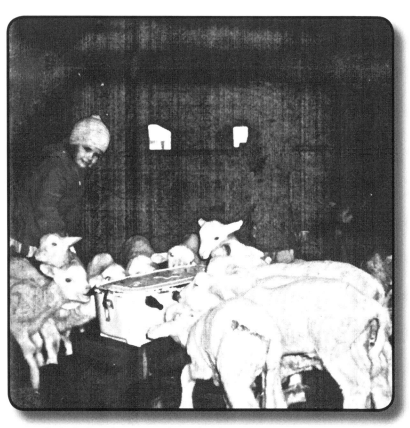

Erin Slivka's first lessons of motherhood were learned in the bum lamb pen, circa 1981.

The Lessons of Motherhood

Many years ago, when I was expecting my first child, I noticed that people began giving me unsolicited advice. Perfect strangers would approach me and tell me things like, "Oh, you better enjoy it now. Pretty soon you won't be getting any sleep!"

When I was expecting the second baby, people would tell my firstborn to enjoy his mommy now because he would soon have competition.

During my third pregnancy, people said, "Oh, pretty soon you'll have two ganging up on the other one!"

By the time number four was causing me to waddle down the grocery store aisle, I was ready to deck the next person who snorted, "Doncha know what causes that?"

However, I found out that people left out some important details about having children. Consequently, I have developed a list of items that I plan to share with the next expectant parents that I see:

- You will have to sneak into the bathroom to eat ice cream at 9 a.m.
- You will never again hear the sermon at church.

- Your children will be clean and well-groomed at all times, but you will go showerless for three days and won't have time to brush your hair.
- The only social outings you will have will be to your friends' children's birthday parties.
- You will be peed on, thrown up on, and have a child wipe his nose upon you, and you will take it all in stride.
- You will be expected to know everything until your child reaches the teen years, at which point it will become obvious that you know nothing.
- You will have to feign excitement at the sight of an enormous black bug perched in the center of your child's tiny hand.
- You will take on the enormous responsibility of teaching small minds the difference between right and wrong.
- You will break the promise you made yourself at age 10 when you swore you would never, ever be like your mother.
- Your only topic of conversation will become your children. Your childless friends will find you incredibly boring.
- In order to decrease the number of times you have to buckle kids into car seats, you will make only one stop when you go to town. If they don't sell it at WalMart, you don't need it that bad, anyway.
- You will discover a new sense of self-importance when you realize you are the only one who knows that the 3:00 a.m. wailing from your child's bedroom means that the snuggly bear has fallen on the floor and the child can't sleep without it.
- You will put on a bathing suit and expose the stretch marks of four pregnancies in public because your children love to go swimming.
- Your will feel your greatest sense of accomplishment at night

when you tuck the last child into bed and sit down for the first time in seven hours.
- You will bake and frost 20 cupcakes, load them up in a minivan, and take them to school once a year for six years for each child.
- Your waistline will never recover from sampling the frosting on said cupcakes, but you will cease to care.
- At 3:00 a.m., you will wish desperately for your baby to be 18 and out of the house so you can get some sleep. At 7:00 a.m., when he flashes you a grin and coos, you wish he would stay a baby forever.

The list could go on and on, but by making it, I have realized something. I will not share it with the next new parents I see, because the value of the list is experiencing it on your own.

The landscape holds reminders that some things, like our heritage and our culture, are not disposable.

A Disposable Society

As I was hauling out the daily bag full of trash yesterday, it occurred to me that we generate a lot of garbage.

A couple of generations ago, families probably did not generate as much garbage in a year as we seem to haul out in a couple of weeks.

The reasons for this phenomenon are numerous. To begin with, we can simply afford to buy more things and thus throw more things away. The things that we buy are not made as well as they were generations ago, so they must be replaced more often. We are less inclined to fix what we have because we have become lazy and would rather just buy a new one.

In addition, we purchase more food from the store, creating more packaging waste. And lately, it seems that everything you need comes in a disposable form.

For instance, in my house at this moment you can find disposable diapers, disposable cups, disposable plates, and disposable silverware. On a daily basis, we use disposable napkins, disposable paper towels, and disposable baby wipes.

The new rage is disposable baby products such as sippy cups, washcloths, and bibs. And I must admit that I have an affection for those disposable kitchen wipes and floor cleaners.

Ten years ago, if you wanted a drink of water, you found a water fountain. Now you pay $1 per bottle for a drink that comes in a disposable bottle.

In our constant quest for convenience, we invented disposable food containers so you can drive up to a window, grab your dinner in Styrofoam containers, and throw them away without washing a thing.

We even have disposable contact lenses and disposable cellular phones.

As a society, we clearly have an affinity for convenience items that require no maintenance. You simply use something and then throw it away.

Unfortunately, products are not the only things in society that are affected by this attitude. As we become accustomed to disposing of things when they no longer meet our needs, we have begun treating people in the same manner.

By reviewing the divorce statistics in this country, it is clear that society considers marriage disposable. When it no longer meets our needs, we discard it.

People are disposable, too, judging from the number of people murdered each day. Even children have become disposable. If a baby does not fit into its parents' plan for life, it can be disposed of and forgotten.

Perhaps it would serve us well to reconsider this disposable mentality and return to a simpler time in which our society did not generate quite so much garbage.

Amusing our Ancestors

If our ancestors could see us now, I think they would be quite amused.

We have spent the last 100 years devising every possible technology to make our lives easier. Because we now have tractors, implements, chemicals, and even GPS, we can now grow more food on less land with fewer people. Between agricultural efficiency, a greater life expectancy, and government policies that encourage food imports from other countries, we now require fewer people on our farms and ranches to grow food for the nation.

Consequently, our cities are growing. More and more people are living in taller and taller buildings, stacking themselves up on top of each other in cities where it is impossible to breathe fresh air.

These people go to work in front of computers that were invented to make our lives easier. They work for eight hours, ride the elevator downstairs, get into their car, and grab dinner from a drive-through window. Then they sit on the couch and watch yet another technological wonder: television.

In the morning, their alarm clock rouses them for another

day of labor.

Not coincidentally, our nation now has a rapidly growing problem with obesity (pun fully intended). So the people are responding. Some are suing the fast-food restaurants for making them fat. Some are reading books that teach them how to eat low-something (low-fat, low-carb, low-taste, etc.). And some are going to the gym.

Can you imagine our ancestors watching as we, along with 37 strangers, wear spandex clothing and pay vast sums of money in order to walk for half an hour and go absolutely nowhere? And then we pay $2 for a bottle of water, get into the elevator, go out the door, find the car, and drive home.

I'm sure the irony of our situation is not lost upon those who came before us. These are people who didn't need alarm clocks. If the sun didn't awaken them, the rooster did.

They didn't have a problem breathing fresh air; it blew freely through the cracks in their walls. They didn't live stacked up in tall buildings. In fact, their nearest neighbor was at least a mile away. They didn't work an eight-hour day. They worked from dawn 'til dusk.

And they certainly didn't read books about the newest fad diet. They thanked the Lord for the food before them and ate their high-cholesterol, high-carbohydrate dinner.

They didn't live quite as long as people tend to live these days, but then they didn't have to take a regimen of 36 pills a day to stay alive, either.

They didn't have to worry much about estate planning because they knew their children would be there to take over when they were gone. They didn't keep in touch with their teenage kids on cellular phones; they kept in touch at the dinner table.

Fortunately, those of us who live in the country are a bit

closer to the old way of life when food came from the field and exercise was just part of the day's work. So, while our forefathers may chuckle as we slide the CD into the player in the tractor, at least they don't see us sweating in spandex.

Finding a new batch of kittens in the straw is a sweet springtime treat.

Barn Cats

Most farm animals know their place in the operation. The horse belongs in the stall. The pigs belong in the trough. The chickens belong on the hood of the newly-washed car. The dog belongs in the back of the pickup. The sheep belong in the neighbor's crop.

These placements are part of farm life as we know it. For the most part, the animals are comfortable with the situation. They accept their position in life, and their owners take it for granted that the animals will not rebel.

All except one species, that is.

Farm animals of the feline sort always seem to be in rebellion. We have assigned them their position on the farm. It's in the barn. We take them there as kittens, feed them there, and assume that they will be happy with their accommodations.

Yeah, right.

Within moments, the cat is tripping you as you enter the house with your arms full of groceries. By the time you set down your bags, the cat is prancing around on the kitchen table like it owns the place.

We've tried to be disciplinarians. We even added the cat's

proper place to its name, possibly thinking that if we call it a "barn cat" it will actually stay in the barn.

This is the only species that requires such a hint. We don't call our bovines "barn cows," but they still know where they belong.

It seems like the harder we try to convince the cat to stay in the barn, the more likely it is to end up on the doorstep.

Farm wives are usually the only people who take pity on the cat.

"It's just lonely," we'll tell our husbands. "I'll just go out and pet it, and then it'll feel better and go back to the barn."

A little pet often ends up in a bowl of milk, perpetuating the problem of the doorstep cat.

The farm wife is more compassionate toward the barn cat because for the woman, the cat represents an object of affection. However, cats represent different things to different members of the family.

To the farm children, the cat is often an object of torture and experimentation. Farm children learn early that cats are expendable, because the Schwan's Man runs over two or three a year. My first lesson on just how expendable cats are came when my mother, decked out in curlers and her ratty bathrobe, stood in the yard with her rifle aimed at the stray cat that had been stealing the dog food. (Stray cats don't merit the same affection from the farm wife because, being wild, they are unable to give and receive affection.)

For farm fathers, the barn cat is usually kicking practice when he trips over it on his way out the door in the morning. However, he tolerates the barn cat as an inexpensive method of rodent control.

For farm dogs, the cats are an endless source of recreation.

Just as each member of the farm family doesn't agree on the

proper treatment of the feline, not all families believe in the same method of treatment for their cats, either. My city friends think that I abuse my cats, but their dog-proof home, endless supply of cat food, frequent snacks of milk (courtesy of small children in the house), and the heat lamp in the winter make them the envy of all the barn cats in the county.

On the opposite end of the spectrum, a neighbor was discussing his cats one day and remarked, "You know, those cats sure do look better now that we started feedin' 'em."

There is general disagreement in the agricultural sector regarding the proper treatment of the barn cat. I have made one observation during my life on the farm, though.

The better you treat your cats, the fewer you will have.

I ended up with 15 cats a few summers ago, so I decided it was time to put a stop to their reproduction. I took my female cats to the vet to get "fixed." While we were there, I invested the extra money for their vaccinations. After all, I didn't want my newly-spayed cats to die of distemper after I had invested money into them.

Of course, within a few weeks the cats that had never strayed beyond the cattle guard at the top of the driveway suddenly disappeared. Now that they were worth some money, they were gone without a trace.

They probably went to live somewhere with worse accommodations. It seems that cats reproduce at alarming rates when they aren't fed properly, not petted, and not vaccinated. They may all be scrawny and have runny eyes, but there are hundreds of them.

No matter which extreme your treatment of barn cats falls under, you can be assured of one thing: agricultural folks understand your predicament.

We know that in the summer you only have two cats, but when the first snowflake flies, every cat ever born on the place comes back. We know that every animal on the place is supposed to have its purpose and its place on the farm. With a little luck, maybe someday we will figure out how to explain that to the barn cat.

Realistic Resolutions

January is such a refreshing time of year on the farm. Why, you only have to spend a few minutes in the 30 mile-per-hour winds pitching feed to your livestock to realize just how refreshing it is.

Traditionally, January is a time of new beginnings and resolutions. Most people resolve to lose that weight they've been promising to lose for 14 years. They may resolve to stop smoking, start exercising, or spend more time with their families.

All of these are, of course, admirable goals. But realistically, they usually amount to nothing more than broken promises by February.

In the spirit of the season, I have decided to make my own realistic resolutions this year:

1. I will cease to spend money on farm cats. Everyone knows that the minute you take Fluffy the cat in to the vet to get vaccinations, Fluffy becomes coyote bait. If you spend more than $20 on a cat, it will position itself squarely in front of the tire of the Schwan's truck the next time it rolls through. Instead of wasting the cash, I will spend nothing on the animals, and they will likely live forever.

2. I will stop "helping" my husband pick out replacement heifers. Last fall, I helpfully persuaded him to ignore his pre-selected list and keep back a nice-looking heifer who turned out to be the calf with the broken leg last spring.
3. I will spend more time teaching my kids about the beauty of the land on which they live. *(Translation: I will let my kids smear garden dirt from head to toe and rinse off with the garden hose while squealing with laughter.)*
4. Instead of complaining about the drought, I will acknowledge its benefits: no mud tracked into the house, lower fuel bills during harvest, lower chemical costs, and more dust on the road to obscure my view of the carloads of tourists driving down the middle of the road.
5. I will understand that it is a genetic predisposition for farmers to wear corresponding hats to dealership open houses, customer appreciation dinners, board meetings, and trips to town for parts.
6. I will stop trying to explain to my city friends why I'm perfectly content to live nearly three hours away from the closest WalMart and why I don't find it problematic to plan two weeks' worth of meals without setting foot in the grocery store.
7. I will stop gasping for air when I'm writing out the fertilizer check.
8. I will attempt to end my jealousy of my husband's dog and his horse, both of whom tend to receive more affection than I do on a daily basis.
9. I will stop wincing when my son pees on the bushes outside and instead thank God that we live in the country where I'm the only one who can see him out my kitchen window.

10. I will stop complaining about the weather. It is easy to see winter as a time of frigid cold and inconveniences, spring as a time of chaotic work and lack of rest, summer as a time of scorching heat and long hours, and fall as a time of too much wind. This year, I will try to see winter as a chance to cuddle up inside with my family, spring as a time of new life, summer as a time of growth, and fall as a time of harvesting the fruits of our labor.

Like all resolutions, these stand a chance of being broken, but hopefully next year will be a fulfilling one for all of us in agriculture.

Chickens

My husband and I have had our share of disagreements through the years, so it's a bit of a relief when we find something we agree on. We have concurred about where to raise our children, compromised on what kind of pizza to order, and found we share a favorable opinion about chickens.

The problem with the latter topic is that we prefer chickens of different types. Shane enjoys his chickens running free on the range, pecking up bugs and decorating the sidewalks with manure.

I prefer my chickens lightly breaded and fried crisp next to a mound of mashed potatoes and gravy.

Where he has acquired his affection for chickens I'll never know, but my husband has always enjoyed the animals. He has built them a living facility which I facetiously refer to as the chicken condo, and they feast on an all-you-can-eat buffet of scraps and grain every day.

One of the most distressing tragedies of the previous year in our household was the day the chickens came up missing. Two were retrieved, but the rest were a tasty snack for the coulee's resident fox.

Since then, we have restocked, and the children are in charge of counting the chickens each day to assure their safety. They feed, water, and entertain the chickens in the morning, and in the evening they shut the door to the condo to keep out any predators.

I remind myself that the responsibility is good for the kids when they come straggling in with chicken manure covering their jeans and shoes. Even two-year-old Matthew takes his job very seriously, and he is never more than two steps behind the others.

Besides, the chickens give the dog some entertainment and some practice when he wants to brush up on his herding skills.

So long as I am not required to kill, pluck, or clean the beasts, I'll live in peace with them, even though the most recent chicken incident made me wonder why we can't just have a goldfish instead.

Last week, the weather was beautifully cool, and the kids returned from their chicken chores to the yard without coming in the house first. I let them play, thinking I could finish the breakfast dishes and have a quiet moment or two before they came back in.

As I watched them out the window, I was counting my blessings that we live in the country where they are able to learn responsibility through chores and have acres to run in the summertime.

I went out to join them in the yard and was helping my youngest into a swing when my daughter said, "Here, Mommy. I'll give you the eggs to take in." I looked around for the egg carton, and when I looked back, she was pulling her hand out of her rear pocket.

"Uh oh, Mommy. I think it cracked."

Trying to maintain my composure, I asked Anna where she put the eggs that she had collected.

"We forgot the carton, so Riley just told us to put them in our pockets," she replied as she went down the slide for the third time.

By the time she reached the bottom of the slide, a yellow streak was following her from her jeans pocket. With a befuddled look, she put her hand to her backside and was disgusted to find that it was goopy and wet.

It was just as I had scraped the last of the shell out of the rear of her pants that I went cold with the realization that if Anna's pockets were full of eggs, her brothers were likely toting eggs as well.

It was difficult not to think of the minimal price of eggs at the grocery store as I scrubbed jeans, washed manure off the unbroken eggs, and dressed the children for the second time that morning. But perhaps it's worth it for the comic relief, as I couldn't keep a straight face when I tried to seriously remind the kids that before they go down the slide next time, they should check their pockets for eggs.

For Matthew, the equipment itch began at the tender age of three.

The Equipment Itch

Each spring, an elaborate ritual begins at our house.

It begins innocently enough as my husband is perusing the latest edition of an ag paper for advertisements. A funny look crosses his face, as if he has an itch that he cannot locate, and a spark flashes in his eyes.

While the cause of the itch differs from year to year, the ritual is consistent. It begins with an advertisement that contains a deal that is just too good to pass up. Regardless of whether or not we need the piece of equipment or vehicle advertised, something about it requires at least a phone call to the owner.

That phone call often leads to a follow-up phone call to clarify a few details, and by then I know that we are traveling down the slippery slope to the bank for an equipment loan.

What started as an itch becomes a full-blown rash that seemingly takes over our lives for a time. A single phone call to the owner of the equipment turns into a flurry of calls as he begins to comparison shop. He consults the Internet, the local equipment dealers, and all of our farming friends and relatives as he contemplates whether or not the original itch-causing advertisement is really the best deal.

Women engage in similar behavior when shopping for shoes. We try on one pair, walk across the mall to another store to try some more, and eventually return to the first store and make a purchase. Fortunately, our purchase rarely involves haggling over an interest rate.

My husband frequently has more than one itch occurring at a time. For example, this spring he is simultaneously dealing on a four-wheeler, a horse, and a tractor to replace the tractor that resulted from a previous itch.

While he contends that he is only investigating possible deals on these items, I am suspicious that he will eventually have so much time invested in one of the itches that he will feel obligated to follow through with a purchase.

Since he knows I have become attuned to his behavior, he now makes some attempt to disguise his intentions. He slips out of the house in the morning and makes phone calls from the shop, trying to act innocent when I see his scribbled notes on scraps of paper.

I know that the rash is becoming quite serious when the phone calls from the dealership begin to increase in frequency.

While a farm wife dreads the look of the itch upon the farmer's face, equipment salesmen can detect the itch from afar and waste no time zeroing in for the sale. That's why I wasn't surprised to find a message on our answering machine from a tractor salesman.

"I heard a rumor that you're thinking about a new tractor. Give me a call!"

And thus the rash continues to run its course.

Laundry Lessons

Since my primary function in life is being a mom, I often lament that the greatest intellectual feat of my day is matching the socks in the laundry basket.

Sure, there are the days when I have to balance checking accounts and project expenses. I sometimes have to write a coherent sentence or two. I can troubleshoot a fever or a rash with the best of them. But when it comes right down to it, I find laundry to be the most consistent source of mental stimulation.

As I was folding tiny socks and Pooh Bear underwear last night, I realized that no one ever taught me how to do laundry. When the time came for me to tackle that task independently, I relied on memories of my mom's methods.

After several years of habitually washing whites first and working toward the darkest loads, with the dirty work clothes last, I questioned my mom about the method. Why is it that we always start with whites? The answer, of course, was that her mom had always done it that way.

When Grandma did the laundry, she used a bucket and a washboard, so doing the whites first made perfect sense. In those days, washing the dark colors first would have meant hauling and heating more water to keep the whites from turning grey.

In today's world of modern conveniences, there is no reason for me to keep washing the whites first in my washing machine. It drains the water with each load. But still, every Saturday I sort the laundry into piles and always carry the whites to the washer first.

I remember swearing as a 12-year-old girl that I would never grow up to be like my mom. I spent the greater part of my teen years promising myself that I would not. But here I am in my 30s, realizing that the women who came before me are ingrained in the rhythm of my daily life.

I look down and see my mother's hands kneading bread dough, pinching it into pieces, and placing the rolls in the stream of sunshine from the window to rise. I answer the phone and hear my Grandma's lyrical hello. I admonish my children and know instinctively that my face looks exactly like my mother's when she spoke the same way to me.

At 15, I would have recognized these influences in horror. Fifteen years later, I marvel that, from many miles away, my relatives are here in my daily routine, and that somehow brings me a quiet comfort.

This knowledge will help me bear the time when my daughters no longer adore me and instead greet my teaching with eye rolls and impatient sighs. I will smile, knowing that one day they will fold my grandchildren's socks and realize that they, too, are part of a circle so strong that it cannot be broken.

Anna, with her dancing eyes and strong spirit, will glance into the mirror and see her mother. Emma Lou, with her patient nature and ready smile, will lovingly rock her babies deep into the night and hum a melody that she subconsciously retrieved from her memories.

Someday, my girls will recognize that my presence in their lives is inescapable. I can only hope that when they do, they will smile.

Looking Good on Paper

Nothing is quite as frightening as trying to make your operation look good on paper.

Bankers and accountants will tell you that you must determine your cost of production, your break-even point, and your net income in order to be an effective manager. They have developed formulas, spreadsheets, and even computer programs to help you accomplish this task.

Unfortunately, it seems that every time we try to plug our numbers into such an equation, we come out with a negative number. That's not very reassuring to us, and I can't imagine that our banker is very impressed, either.

Since we have been operating for several years and we have not yet had to file for bankruptcy, I am under the assumption that we are, at some point, making some money. It just never quite pencils out that way. Somehow, agriculture just doesn't translate into neat little formulas and spreadsheets.

One of the major problems is trying to estimate expenses and income over a one-year period when the professional economists can't even tell you what the markets will do next week.

Estimating your income is a bit risky when one year you sell wheat for $8 and in subsequent years you are happy to get half of that.

It is also impossible to predict expenses in agricultural operations. Depending on which diseases, weeds, and pests are thriving in a particular year, chemical costs can vary enormously. If a cattle disease strikes, vet bills can skyrocket.

You are also at the mercy of an unstable fuel market, which can have devastating effects on your bottom line.

And, of course, there is also the catastrophic tractor breakdown that sends your neatly-figured budget spiraling down into the circular file.

Based on these variables, you can choose one of two approaches. You can either be an optimist and estimate that your income will exceed your expenses, or you can be a pessimist and assume that disaster will strike your operation and you will depend upon disaster relief funds to keep your operation financially sound.

I imagine that after 25 years of involvement in production agriculture, it is a bit easier to make estimates on your bottom line. I am still dubious that anyone can ever fill out one of those forms with great accuracy, but having a 20-year average to operate from must surely make the process a little less painful.

In the meantime, our relatively recent adventure into the world of cattle ranching is still struggling to look good on paper. Let's hope that the banker is more of an optimist than a pessimist.

Modified Housekeeping

As I was attempting to tidy up the house the other day, I looked around and realized that not every woman lives like this. I belong to a special sisterhood of women who have a jug full of colostrum in the freezer. I didn't bat an eye when my husband came in from the barn and set a large pitcher of frothy yellow milk next to the coffee pot on my kitchen counter.

After all, this is a house that usually has an enormous rubber nipple air drying next to the sink and a bottle of Banamine in the butter compartment. When doing the supper dishes, I shove aside the vaccine gun and stomach tube that are drying in the dish rack.

In the bottom of the washing machine, among the Legos and hair barrettes, I frequently find a copious amount of straw beneath the load of jeans. My dryer's lint trap is accustomed to collecting those little green castration rings. I'm always sure to check the shirt pockets before they go into the washer, though. The calving book must be removed and protected as if it were the Holy Grail.

While it's not unusual to see a few pairs of mucky coveralls in the porch, the worst of the dirty laundry often hangs on the deck railing, waiting to be hosed off before it enters the house. When the stench of muddy manure emanating from the muck boots in the porch starts to overpower the smell of fresh bread baking in the kitchen, I know it's time to throw open the windows and air it out, even if it's 20 degrees outside.

Since I've lived on a ranch nearly my entire life, I am accustomed to this modified version of housekeeping. I know the secrets to removing ear tag marker from the wall and newborn calf slime from jeans.

But to a person not used to this particular lifestyle, our habits may raise an eyebrow. The FedEx deliveryman is probably baffled at the collection of dried placentas that our dog has stashed in the corner of our yard to be gnawed on like jerky. We know it's a special treat for him because he keeps it right next to the deer leg that he salvaged last fall after hunting season.

We don't often leave the place during this time of year, and when we do, it's difficult to switch out of animal husbandry mode. In honor of the rare trip past the end of the driveway, I try to replace my manure splattered sweatshirt with a nice clean version, and I usually substitute a clean ball cap for my wool ear flapper cap that is a critical component of my calving uniform. With a nod to my delicate feminine nature, I even brush the ground feed out of my ponytail and wash my face.

More than once this season I have caught myself heading to the car with a sorting stick in my hand. It has become second nature to grab a stick on the way out the door to fend off any overprotective mama cows who see a bullseye on my rear. And although I try to leave all thoughts of cattle behind me, the other day I caught myself describing my sick child as "off her feed."

Rejoining humanity is difficult when we have spent weeks living with cows. Being a multitasker, I cannot stand at the kitchen sink without looking out the window for that kinked tail that indicates an imminent birth. I can fall asleep sitting upright when I'm waiting for the clock to roll around for my next check, but I jolt awake when I hear the telltale moo of a new mother, especially if it's a heifer who thinks a UFO just dropped an alien being in the pen with her as she meets her calf for the first time.

Eventually, life will return to a more normal state. The calving paraphernalia will be shelved, my counters will be cleared, the muck will dry up, and the cows will go to pasture.

But until then, if you happen to drop by, please don't be offended by my shoddy housekeeping. I promise that I used pasteurized milk in the bread I just made, and I was sure to thoroughly wash my hands before I kneaded the dough.

Suffolk, Montana

The Price of Progress

A local journalist recently wrote a column about the travesty of Montana's lack of progress. She had recently taken a trip to another state, and on that trip she realized just how many conveniences Montana lacks. She was disgusted that Montana lags behind the "rest of the world." In her view, Montana has not made appropriate levels of progress in the past 50 years.

I agree. If Montana wants to keep up with the "rest of the world," it has some catching up to do. But I don't think that Montana wants to keep up. I think Montanans are pretty happy with their way of life.

What gives me that idea?

First of all, anyone who lives in a small town knows that change is a difficult task to accomplish. Traditions run so deep in Montana that to change them is nearly impossible. In fact, it is rarely even considered.

These traditions can be matters as significant as the annual Fourth of July parade or details as small as the brand of coffee served at the local café. To change anything about either one would be as shocking as if Farmer Joe, whose family has driven Fords for three generations, would suddenly drive down Main

Street in a Chevy.

I think Montanans are content with the slower pace of life here.

I had a good friend in college who hailed from Los Angeles. Watching her was like watching the Tasmanian Devil on Saturday morning cartoons. She walked fast, talked fast, drove fast, ate fast, and thought fast. When we would tease her about it, she would giggle. She said that life moves so quickly in L.A. that a person just acclimates to that fast pace. She said that Montana was very relaxing for her because we all do things so slowly.

The results of that slower pace are far-reaching. We tend to think before we speak and act. Perhaps that is one reason why our crime rate is lower. We drive more cautiously, which results in rude gestures and honking in town, but it allows us to notice small details about where we live. We stop in the middle of the road and talk to the neighbor. That would never happen in L.A.

The slower pace also results in more patience. Because we move a bit more slowly, we are also used to waiting for things. We wait to be served in the local café while the waitress carries on a detailed conversation with our neighbor concerning the weather. We wait for the grocery clerk to help someone to the car with their bags before she comes back to ring up our purchases. We're just used to waiting, so it doesn't aggravate us like it does people who come from a faster-paced life.

There is a special quality about life in Montana that I just can't put my finger on. It's comfortable. It's friendly. It's home.

If that quality was created because Montana failed to keep up with the "rest of the world," I think we should be anti-progressive. I like the fact that I can step into a café in any small town in this state and feel just like I did when I sat in our local café as a child. I hear the same conversations about weather, prices, farming techniques, and the local ball team. The food tastes real, and I enjoy waiting for it. I don't need the drive-through of a national chain store.

We have all complained about our small towns. We complain that there isn't enough shopping and that we need a Wal-Mart a little closer to home. But do we really want that Wal-Mart and the faster pace that those conveniences bring?

I don't think so. And, as I have often told critics of Montana, if you don't like this state, there are plenty of others to choose from.

A father and his flock at a popular vacation destination

A Working Vacation

This summer has danced by in a flurry of motion, and with the first day of school taunting us on the calendar, we decided that it was time to get away from it all.

Now, for some families, that might mean a relaxing vacation, perhaps on a beach with plenty of sun, no worries, no responsibilities, and room service. For kids, it means a trip to Disneyland or a water park. But since a week without work would seem so strange to us, we set out to visit our friends who own a dairy, where the chores never end and an extra set of hands is always appreciated.

While the prospect of a working vacation didn't really bother me, the fact that the dairy is located in Wisconsin caused some concern. I'm not sure what possessed me to agree to a 1,150 mile trip across four states with four children and a farmer. Perhaps it was the fact that I only had 24 hours' notice of our departure, and I was so busy packing that I didn't consider all the possibilities of the journey.

The risk, of course, involves traveling with a farmer. I can fully prepare for the trip with the children, providing snacks, drinks, and activities to keep them entertained. The farmer, however, has trouble keeping his focus on the miles ahead. He is distracted by tractors flaunting "For Sale" signs, undeterred by the fact that stopping the vehicle means waking a sleeping child and setting off the whining mechanism.

Our children are accustomed to traveling with their father. They know that they shouldn't ask if we can stop and use the bathroom. Instead, they ask how much farther it is until the next tractor dealership, since that always means a stop to use the facilities. This vacation was considered particularly successful since I only had to enter one John Deere store to remind my husband of my "One Hour in the Parking Lot with Four Bored Children" rule.

Although the magnetic force of implement dealerships is a road hazard for our family trips, I also hold my breath every time we exit the freeway, and I knew we were in trouble when we did so in North Dakota and ended up on a dirt road where I could have rolled down my window and touched the corn. These situations make me nervous not only because my husband has been known to randomly leave the car and wander around someone else's crop, but also because of the ironic fact that we have some of the latest GPS technology installed in our farm equipment, but we have yet to spend $100 to install such a device in our vehicle. Instead, we rely on maps to navigate our way across strange places, arguing about exits and straining to see signs and change lanes at the last second.

That explains how we missed the bypass around the Twin Cities in Minnesota and ended up driving straight through the heart of Minneapolis, giving the children the double thrill of seeing skyscrapers up close and seeing their father hyperventilating in the passenger seat as we navigated through city traffic. Hours earlier, we had been marveling at the sight of the World's

Largest Holstein Cow sculpture, and we were suddenly in a maze of roads that crisscross their way through a metropolitan area whose drivers are unprepared for the likes of us.

We escaped safely and eventually found ourselves in a land so different from our own that the kids thought we must be in a foreign country with green grass in August, pavement on every road, bountiful flower gardens, towering corn, trees on every hilltop, ferns and vines in the road ditches, and orange tractors.

After many miles of soaking in the sights of farmland that receives actual precipitation on a regular basis, we arrived at our destination. Although we had traveled for two days and were standing on ground that was quite unfamiliar, we instantly felt at home as we greeted our friends and were reminded of the universal qualities of people involved in agriculture. For a few days, we immersed ourselves in their world, putting up silage, feeding calves, fixing flat tires, welding equipment, and feeding the crew. While we broke my husband's rule of avoiding any activity involving the word "tour," we found that our favorite part of the trip was stepping into the experiences on the farm.

I am a little bit hesitant to send my kids back to school for fear they will tell their class, "On my summer vacation, I went to Wisconsin and saw a big lake of manure on a dairy farm!" On the other hand, hearing about their favorite activities makes me smile. One child loved seeing the baby pigs on a neighbor's hog farm. Another loved playing with the toy tractors in the calf barn. Our oldest girl liked the night we were target shooting in the corn field with the guys from the silage crew.

It wasn't Disneyland, but I have a feeling that a trip to a Wisconsin dairy farm is something our kids will never forget.

The Story of My Life

Two of the most frequently asked questions that I field (besides "What's for supper?") are "How do you find the time to write?" and "How do you find topics for your column?"

Since the answers to those questions were eluding me, I decided to document my own process of creating a column, and in retrospect I believe that the answers to those two common questions are intricately related.

The process begins when I check my email after the kids have gone to bed and find a message from Gail, the lovely lady who beams over a reminder from Bismarck. Not only does she give me a yearly schedule of paper deadlines, but she regularly gives me a one-day warning when a column is due. I usually drift off to sleep while attempting to decide upon a column topic.

The next day, I think of no fewer than 20 possibilities of column topics. I fail to write about any of them because I am usually busy with other tasks, and by the time I remember to jot down the idea, I have forgotten it.

That evening, I check my email after the kids have gone to bed and find Gail's second reminder that my column was due. I stare at the blank document screen for a considerable amount of time before deciding that my brain will be much fresher in the morning. I drift off to sleep, still attempting to decide upon a column topic.

The next morning, I sit down to the computer, intent upon writing a complete column. Before I am two sentences in, the youngest child is awake and requesting breakfast. The following hour passes in a frenzy of breakfast making, coffee drinking, dish washing, spilled milk cleaning, library book finding, hair combing, teeth brushing, catching the bus chaos. With the older kids out the door, it's time to prepare the little ones for the day and do the morning chores.

On the way back from the barn, I am mentally planning the day's menus. After the calf bottles are washed and the little ones are settled with their toys for a few seconds, I remember that a bill needs paid before the mail goes out today, so I write a check and make a quick trip up to the mailbox.

Home again, I check the grain markets momentarily before I sit down to tackle the unfinished (and now overdue) column. By this time, the little ones need a story or two before they transition to their next activity of the day. I bundle them up and shoo them outside with warnings of staying out the mud puddles, and I again sneak off to the computer to work on the task at hand.

Two paragraphs in, the back door slams and my husband, having completed the morning feeding, requests the number of the calf we doctored two days ago. Retrieving that information for him, I return to the computer. A few more sentences are gained before the phone rings. The John Deere sales representative would like to speak to my husband, which arouses my suspicion.

After a phone call from the vet and an additional call from the Case IH sales rep, I am not only suspicious, but also off track.

By the time I regain my train of thought, my husband is at the door again requesting a weather update and the checkbook.

The kids arrive at the door on his heels with an announcement. "Sorry, Mom. We forgot to stay out of the mud."

Once I clean up the muddy tracks in the kitchen, I give up on the column writing and prepare lunch. Lunch and dishes conquered, I check my email to find a note from Gail politely requesting my column so she can meet her own deadline. After making a few quick phone calls for church activities, I return to the column. I have now forgotten the column idea altogether.

My husband engages me in a conversation about loan balances. Clearly under the influence of tractor sales reps, he attempts to convince me about the merits of leasing, which evolves into a conversation about our long term goals and the future of agriculture.

Now thoroughly distracted, I do some early supper preparations and then go to the end of the driveway to retrieve the older kids from the school bus. The after school chaos of hugs, snacks, chores, and homework ensues. I make a partial grocery list while finishing supper preparations and coaching the sentence diagramming that is taking place at my kitchen table.

On a good day, I am able to find a 30 minute window of time after the kids are in bed during which I have no interruptions and can complete the column.

On a bad day, I am called upon to help "for a minute" and find myself sorting pairs in the sleet, splinting a calf's broken leg, and wondering why I didn't find a nice accountant to marry.

I seldom become terribly frustrated during the column writing process, though. I realize that all of those interruptions throughout my day are the real topics of my columns. Without all those experiences, I really wouldn't have anything to share with the faithful readers of this column. And I suspect that being able to relate to those daily experiences is the real reason that people read the column to begin with.

The truth of the matter is that as soon as I find that I have plenty of time to write, I will also find that I don't have anything to write about. Interruptions, it would seem, are truly the story of my life.

Tending Lucille's Garden

Spring has always been my favorite time of year.

One morning you wake up and realize that the grass is green, and the sun has a new warmth that you haven't felt since the onset of fall.

The sun's rays melt the frost from the ground, which begins to emit an earthy smell that every farmer and gardener loves. It signifies that planting time is approaching, and for those who love to sow seeds, that smell holds the promise of a new season of nurturing new life.

Last week I threw open the windows and let that smell permeate through the house. But a touch of sadness crept into the joy spring always brings to my heart. There is one less gardener in the world this spring.

Lucille's garden was amazing, I'm told. I was never able to see it firsthand; by the time I met her, the years were taking their toll on her once sharp mind. That fall she moved to town, away from the vast plot of land that was once her precious garden.

As Alzheimer's ran its course, family and friends watched in sorrow as the light in her eyes began to dim. The present was confusing and overwhelming, but one subject would light the fire in her eyes once more.

"Do you remember your garden, Grandma?"

"Well, yes," she would reply, looking up as if you were an idiot to think she could forget. "I hope somebody will weed it for me while I'm gone. When I went out to Washington, the weeds had taken over by the time I got back."

We promised to take care of it for her, but she knew the truth. We could never do it right; only she had the magical touch that could make a garden thrive.

When family and friends gathered to say a final goodbye in February, many memories of Lucille surfaced. A granddaughter remembered how she would nap in her chair while holding a newspaper. A grandson remembered that instant pudding always tasted better at Grandma's house.

Almost everyone spoke of Lucille's garden.

"How she could make things grow out of that pile of rocks she called a garden, I'll never know," marveled one friend.

"She made the best pickles," remarked another.

"She knew all the best gardening tricks," said yet another. "Did you know that scattering human hair around the garden will keep the deer away?"

Lucille grew wonderful vegetables, without a doubt. But more than that, she was a master nurturer. Just as she nurtured her garden, she nurtured a family that has sprawled out across the country like her cucumber vines sprawled across the garden.

Children and grandchildren and great-grandchildren remain as a testament to Lucille's life and her ability to nurture. As I look down at my baby boy, who is smelling that earthy smell for the first time this spring, the joy returns to my heart.

We will tend to your garden, Lucille.

Foodies

$\mathcal{I}$n the 30 years that have passed since I learned to read, I have spent a good percentage of time with my nose in a book. I also enjoy reading newspapers, magazines, labels, instructions, farm publications, and even the occasional piece of junk mail. Since subscribing to satellite internet several years ago, my horizons have widened to the nearly limitless landscape of the web, where I can follow blogs, discussion boards, market news, and the comment sections of local newspapers.

While the access to enormous amounts of information has certainly expanded my knowledge on many topics, it has also reinforced my belief that common sense among humans is nearly extinct.

That conclusion is the result of an entire year of reading the responses of the general public to the film *Food, Inc.*, which seems to have become the authoritative source of information about agriculture for everyone expect those of us who are actually involved in food production. If I had a buck for every Twitter or Facebook status update that said, "Eeewww! I just watched *Food, Inc.*, and I'll never eat another hamburger," I would be able to pay off our loans and have money left over for a nice steak dinner.

In that regard, the film's creators have accomplished their goal. They have built upon the growing movement of public opinion that views agriculture in a negative light, vilifying any agricultural operation that dares to seek a profit, grow beyond a few acres, feed grain to livestock, or house an animal in conditions that fall short of the Hilton. Representing the entire beef industry with repetitious footage of a downer dairy cow makes for a good scare tactic. Representing the poultry industry with an interview of a chicken farmer who is going broke makes for a good emotional appeal. But the film is just the beginning of the *Food, Inc.* story.

The official website of the film is less subtle in its message. It blatantly states that "industrial farms" are raising animals under inhumane conditions as well as causing pollution, putting their workers in danger, contributing unsafe food to the food system, and causing global warming. It cites pesticides as the cause of autism. Most notably, it blames the United States' agricultural policies for starving 1 billion people worldwide and sickening 76 million Americans every year.

One columnist's summary of the issue is that "agribusiness promotes an unhealthy diet."

The obvious responses of a person not exposed to the truth about agriculture would be fear and anger, and scared and angry people seek protection. They expect the government to provide that protection, and they call for it in the form of increased regulations and more government involvement in food production. They want restaurants to stop serving unhealthy food. They want schools to go organic. They want to have a say in how crops and livestock are produced, even though they don't know the first thing about farming. In their minds, they know better how to produce a quality food product than do those of us who have been growing food for our entire lives.

The "foodies" of our society, those who are preoccupied with the eating, cooking, and discussion of food, are now delving into the topic of food production. They tout organic food, sustainable agriculture, grassfed beef, and slow food without actually knowing the facts regarding any of those issues. A generation or two ago, foodies were few and far between because people were busy working up an appetite and trying to put food on the table. They worked hard enough to appreciate whatever was on their plate, and they were hungry enough to eat it.

Now, due to the industrialization of our country that resulted in a system in which not everyone has to grow his own food, people have just enough time on their hands to watch movies like Food, Inc., but not enough time to research the truth.

I read the writings of these people nearly every day, and I always return to one argument: people can choose. We cannot blame agribusiness for the diabetes epidemic. We cannot blame the ag industry for global hunger, and we cannot blame Monsanto for my son's autism. We cannot blame fast food restaurants for childhood obesity. We cannot blame tobacco farmers for lung cancer deaths.

People have choices, and the responsibility to make healthy choices belongs to an individual, not to a government or an industry. Unfortunately, I can count on one hand the number of places I have read a common sense statement like that. Maybe I'll just have to keep reading.

The Importance of Being Specific

A Snowy Thursday in January

Dear Diary,

The walls are closing in. Snow has besieged our area, making even a simple trip to the grocery store a challenge. I'm developing a critical case of cabin fever. My husband suggests AI school. I'm open to any idea that gets me out of the house for a few days.

A Snowy Friday in February

Dear Diary,

Today I scaled the glacier in our calving lot at least four times while trying to outwit a three-year-old cow who is in much better shape than I'm in, even though she calved about 20 minutes ago and has yet to shed the placenta. I desperately need a getaway. AI school is sounding like a promising possibility. It will last four days, and since it's tax deductible, I'll have a great response if my husband complains about the cost.

A Windy Wednesday in March

Dear Diary,

Calving has gone splendidly well. The weather is beautiful.

We've spent most of our time outdoors lately, and the sunshine is good soul therapy. I received word that I've been accepted to AI school. I'm having doubts about this idea. Who's going to watch the kids? With this early spring, we'll be seeding for sure. And I don't really need to get away from it all now that the snow is melted and the sun is shining. I wonder if it's too late to back out.

A Dry Day in April

Dear Diary,

My husband says I need to follow through with this AI school proposition. I'm not keen on the idea. Why would I want to spend four days with my hand in the nether regions of a cow? This is not at all the getaway I had envisioned.

AI School, Day 1

Dear Diary,

I don't think this is my calling. The morning went well, and I seemed to catch on to the concepts of AI with no trouble. However, the reality of the matter struck when the instructor plopped on the table the reproductive tracts from several formerly alive cows. Armed with an AI gun, we were each commanded to place the instrument through the cervix of the mass of tissue in front of us. And that was not the grossest part of the day.

After we were deemed capable of wielding an AI gun without major injury to the formerly alive cows' reproductive tracts, we were set loose on the real thing. I've had plenty of humbling experiences in my life, but none quite measure up to going armpit deep in a cow's rectum and cleaning her out so I could feel for a cervix. I was feeling rather inept and embarrassed until the guy next to me declared that his cow clearly didn't have a cervix, but he was pretty sure he was in deep enough to feel her tongue.

When I called home, I didn't have much to report except for the confession that I had been peed on twice and I was sure glad I had grabbed my husband's coveralls and not my own for this excursion.

AI School, Day 2

Remarkably, every member of our class of 14 returned today. I was disappointed that there weren't doughnuts. I only thought that the worst of it was over, but I was sadly mistaken. Today I felt like I was back in junior high, possessing that fumbling awkwardness that every adolescent experiences. Let me just tell you that retrieving a semen straw out of a tank of liquid nitrogen is not as easy as it sounds. Several dropped straws and nitrogen blisters later, I can attest that I will never get this right. So much for my big getaway. I'm having trouble sleeping because I keep dreaming that I'm reaching for a cervix in a vast cavern of darkness. Sadly, I know that when I get up tomorrow morning, that's exactly what I'm going to be doing.

AI School, Day 3

Today I learned that cows can blow themselves up like beach balls internally. The first cow I encountered could have housed a small brass quartet inside. When I employed the techniques for disbursing the air, she cooperated willingly, blowing it all back into my face only to suck it all back in again and clamp down on my arm until the circulation was completely cut off. I reminded myself that I was paying for this privilege and then dreamed of the life I would have had if I had only found a nice accountant to marry. Only when I was on the brink of giving up did I finally find success today, and I almost felt a twinge of confidence after threading the cervixes of six consecutive cows. I was further bolstered by the confessions of two of my classmates that on the first day of school, their arms had entered the wrong opening of their cows. They only realized their mistake after finding the cervix a little too quickly and accurately. Why is it that we feel so much better about ourselves when someone else screws up?

AI School, Day 4

I retrieved a semen straw from the AI tank without pain, injury, or embarrassment today. I effectively deflated an old Hereford long enough to find her cervix. I avoided the urine stream of every cow, and I was not the unlucky student who pulled the practical joke golf ball from the rectum of the first cow of the day. All in all, it was a productive morning. I'm still a little disappointed that there weren't doughnuts, but thanks to the stellar patience and infinite wisdom of the instructors, I have much more knowledge of the business today than I did last Friday. I discovered that I'm not too old to learn something new. But next time I tell my husband that I need a getaway, I'll probably be a little more specific.

 Bad News

This time of year, I cringe when I see my spouse during the daylight hours.

Let me clarify that.

My negative reaction is not due to the fact that he desperately needs a haircut and a shave. It's not because he has had too much contact with the back end of a cow. I don't even mind that he is wearing the same clothes he has worn for the past three days.

It's not that I don't want to see my spouse; it's just that during this time of year, his presence at the house in the daylight can only mean bad news.

With the emergence of the wild oats and the convergence of chemical salesmen buying drinks at the local bar, we can be sure that spring is upon us. And during springtime on the farm, the male half of the partnership is seldom seen during the daylight.

Between feeding, calving, fencing, branding, and seeding, most farmers and ranchers these days are too busy to eat, so when I see my husband trudging toward the house, I automatically shudder as I imagine the bad news he must be bearing.

It is quite possible that something is broken. After a long winter of disuse, the farming equipment can usually be counted upon to have some kind of a hiccup.

If that's not the case, it could be a casualty in the calving department. Since only a handful of cows remain to calve, they have fallen down on the priority list, which heightens the possibility of a dead calf and a flurry of phone calls to locate a graft candidate. It could be that he is coming in for lunch at 3:30 p.m., having finally finished the chores he set out to do at 6 a.m.

One of the scariest sights I have seen recently is that of my husband walking in the door wearing a new cap and carrying a new coffee mug. Normally, I would not make a fuss over his buying a new cap and mug, but when he acquires them at a bull sale, my heart begins to quicken. Just how much money did he have to spend in order to come home with both a cap and a coffee mug? I have learned the hard way that complementary caps are the most expensive headgear available.

No matter what the scenario, farm wives must follow a few simple rules to survive springtime.

First, they must never, ever ask the farmer, "How was your day?" That is simply asking for trouble. Second, wives should always have something in the refrigerator that can be microwaved in seconds when the farmer enters the house at 11 p.m. only to fall into bed too tired to eat it after all. Another safety precaution is to say as little as possible. Your husband doesn't need anything else to think about, and chances are he's not listening, anyway.

Finally, for goodness' sake, don't break anything, and if you do, hide it. Under no circumstances does a farmer want to come in during the stressful springtime rush and discover work to do in the house.

Following these bits of advice just might take some of the stress out of springtime and allow both you and your spouse to see the renewal of the land in a brighter light.

Hometown Heritage

This summer, my hometown is celebrating a significant milestone: the 100th anniversary of the school.

The resilience of the original brick building is not a small feat considering its location. Judith Gap is notorious for its fierce winter weather and its incessant wind, a feature now famous due to the wind farm that carries the town's name.

But the celebration is not so much about the building inscribed with the date of 1911. The celebration is about the generations of people who have walked the halls there and learned much more than just reading, writing, and arithmetic.

Following my grandfather's footsteps down the halls of that school, I learned how to succeed and to fail. I learned about defeat over and over on the basketball court that was so narrow that a baseline three pointer must be shot from the bleachers. Underneath the squeaky gym floor in the school shop, I learned to offset my lack of athletic ability by succeeding in agricultural education. We tore apart motors, welded a uniform bead, built livestock feeders, ran profit and loss spreadsheets, and learned livestock judging skills that have served me well long after my free throw shooting ability has been forgotten.

On the top floor, I learned problem solving skills from the finest math teacher that ever conducted a class. Missing a problem meant disappointing her, and I worked diligently to keep that from happening. Across the hall, an equally demanding English teacher honed my love of reading into the ability to write. Her insistence on perfection and massive amounts of homework were constant irritants throughout high school, and I was proud to carry on her tradition of high expectations when I became an English teacher a few years later.

A walk through the school elicits a flood of memories from Kindergarten graduation, when my new piece of chalk broke in half and resulted in tears, to those years in high school when I thought I knew much more than I ever did.

I can still smell the ink of the ditto machine in the elementary hallway and see its purple stains on the timed multiplication tests that made me sweat so much. The formaldehyde scent of the science room takes me back to frog dissections and intense arguments with the teacher over the best way to solve a physics problem. The sound of the lockers slamming makes me reminisce about the friendships that seemed so solid that have faded away through the years as we have all drifted apart, away from that common foundation that began in the little brick school.

One thing has remained consistent through all the years of change, however. The town is just as proud of that building and its occupants as it was when it was built 100 years ago.

And I will always appreciate the lessons learned in my 12 years of education there. Like everyone else who has walked those halls, I am proud of my roots and how they have shaped me. I'm not ashamed that I graduated in a class of seven. I'm glad our town's phone numbers only take up one page in a very small book.

And even though it's on a different page in that phone book, I'm glad that our phone number is listed in a similar section just a few pages away. My kids, too, will have the memories of moving through the rooms of an old brick school building and knowing that they are a source of pride for the entire community. It's a foundation worth celebrating.

Scour Guard in My Stocking

The time of year is upon us when people divide into two camps: those who love to shop and those who criticize the commercialism of Christmas in order to get out of shopping.

I have, on occasion, considered the possibility that we have lost the true meaning of Christmas amidst the glitzy fuchsia garland and screeching salespeople that mark the beginning of the holiday season. That possibility grew more real in my mind this year as I tried to complete my Christmas shopping, which has become more of a challenge in recent years.

It didn't take many forays into the shopping mall with three preschoolers to convince me that mail order is the way to go. Besides, shopping by mail gives me the luxury of looking over catalogs and wondering who actually buys this stuff. I have found items that range from truly amazing to just plain stupid.

In one catalog, I found a $75 CD player shaped like a Volkswagen Beetle with speakers in its wheels. Another catalog sported gold earrings shaped like antlers and a phone that could easily double as a duck decoy.

Other options included a contraption on wheels that holds up to 24 fishing rods. This made me wonder, are there really people who possess 24 fishing rods? Astonishing.

For those snackers on your list, $29.99 will buy you a S'mores maker complete with ceramic ramekins to hold the marshmallows. I'm not sure what a ramekin is, but I had trouble envisioning our family gathered 'round the gas burner holding our ramekins and singing campfire songs.

In the same catalog, I found that I could send my loved one a $40 carrot cake or a $200 robotic vacuum cleaner. Quite a selection.

Another intriguing option was to send two pounds of meatballs for a mere $19.99. I guess next year we should bypass the cattle buyers and just butcher our steers, grind them up, roll them into little balls, and ship 'em on dry ice.

For those who really "have it all," catalogs offer gifts like an Elvis ornament, a pair of battery-heated socks (which would make a great gift for a future groom), or a "Smart Mug" which plugs into your 12-volt and keeps your coffee at a constant 160 degrees.

A great option for the farmer in your life is the alarm clock that projects on the wall not only the time, but also the current outdoor temperature. No need to get out of bed and walk to the window to see the thermometer; now you need not even roll over.

For the pet-lover, the personalized pet bandana is definitely the way to go. Of course, you could splurge and purchase the "Bow Lingual" device that claims to translate your dog's every sound into words. I'm not sure about you, but I don't think I want to know what my dog is saying about me. (By the way, this device apparently has no problem with purebred language, but it may not be able to decipher those lower-class mixed breed barks.)

If a person was still stumped for a gift, they might consider the ever-popular funnel that allows you to save the last tiny bit of ketchup from the old bottle and transfer it over to the new container. You can also buy a musical jukebox alarm clock with dancing lights, a Santa shower curtain, or pink flamingos for the yard.

Buying for that special guy is no longer a problem. Now available is a cab enclosure for an ATV that looks like a pop-up tent riding around on your four-wheeler. Those with bathroom decorating dilemmas can choose the acrylic toilet seat with fishing lures or bullets inside.

I must admit that my favorite holiday flyer came from Big R. I can just see the adoring look on my husband's face when he opens up that package containing the grease gun and the scour guard.

After looking through all the catalogs, it is certainly easy to think that there is something wrong with a society in which we have so many material possessions that we have to buy each other useless, frivolous items to celebrate the birth of Christ.

However, I then consider the first gifts given to Jesus. Here was a child whose parents did not even have clothing for a baby, and what did the Wiseman apparently bring Him? Frankincense, gold, and myrrh. These are not exactly practical items.

So what made them the perfect Christmas gifts?

The intent behind Christmas gift giving goes beyond finding a practical gift or spending enough money that the recipient will be impressed. The intent is to give of oneself to another. For some people, that means buying a Volkswagen CD player. For others, it's a grease gun. And that's okay.

Battling Bovine Racism

Occasionally, I take on extra freelance writing assignments. I'm not quite sure what possesses me to agree to additional work; perhaps it can be explained by the fact that the most intellectually stimulating portion of my day is oftentimes matching the socks in the bottom of the laundry basket. Maybe I enjoy the challenge of conducting interviews over the phone while my five-year-old pretends he's an airplane and my toddler attempts death-defying acts of courage atop of the kitchen table.

Whatever the reason, I recently agreed to write an article about the culling strategies of cattle producers.

I confess that I had never researched any scientific strategies involved with culling cows. While I was growing up on a ranch, our culling system seemed quite simple to me.

The cattle or sheep with the big red Xs painted deliberately on their heads were the ones to take the fateful trip down the chute to the stock trailer. The offending critters earned their Xs in a variety of ways. Of course, there were the standard culls: those who failed to calve, those who prolapsed, or those who had milk supply deficiencies.

The other category of culls involved those animals who, at

some point during the calving or lambing season, had earned a revered name involving language not fit for a family publication. The rejected animals had usually done some kind of bodily harm to either their owner or their offspring.

Of course, age and disease were other reasons to send a cow to the auction yards.

In addition to the culling of the cows, decisions were made each year about heifers as well. This is where I fit in to the culling system. I operated on a simple system identified as the cuteness test. If Dad indicated he would keep 10 heifers that year, I would simply form a hierarchy of the cutest heifer calves in the herd. The rest were culls.

It wasn't that I was ignorant of desirable traits in replacement heifers. I performed decently in my FFA livestock judging competitions. Unfortunately, my training was quite forgotten when I gazed into the heavily-lashed eyes of a brockle-faced heifer calf with a white-tipped tail.

My culling strategy, then, was to apply some sort of spell upon the better judgment of my father, who was more apt to choose his replacements based on maternal characteristics or growth potential.

I am proud to say that my strategic culling practices are still intact. When my husband branded his first crop of calves and was deliberating about possible replacement candidates, I began my lobby for the cutest calves. The neighbor who helped us brand rolled his eyes as I made my picks. A well-sprung, feminine-looking pure black heifer received a thumbs down while a doe-eyed, reddish brown heifer passed the test.

I'm not intentionally trying to sabotage our cattle operation, but I also don't want our herd to look like nearly all the others in the county: pure black with no variety.

"It's racism," I told my husband on a recent drive to survey the cows. "The black cows are receiving preferential treatment for no other reason than the color of their skin."

It remains my duty, then, to ignore the research in the article I am writing in a valiant effort to save the colorfulness of the landscape. I will fight to keep the brockle-faced, the mottled-colored, and the overall cute calves from the culling gate.

And somehow, I know that there are other farm wives out there who are also engaging in the greatest culling strategy of all.

Saving Montana

Ten years ago, I wrote a column about the creation of the Missouri River Breaks National Monument. At the time, I was frustrated that a group of politicians could come float the river for a few hours and then declare that it needed to be protected so more tourists could enjoy it in the future.

Such statements are an insult to the ranchers whose families have been good stewards to the land for generations. For politicians, environmental stewardship is a buzzword thrown about during election years to appeal to the green voters. For agricultural families, environmental stewardship is taking care of the land in order to make a living and pass it along to so the next generation can do the same.

Ten years ago, I mocked those who were touting the economic benefits that tourism would bring to this area upon the creation of the national monument. Proponents of the monument were adamant that our small community's reliance on agriculture would doom it and that capitalizing on tourism would be our saving grace.

I proposed to abandon our farming operation and instead set up a jackalope sales stand along my driveway, which happens to be on a popular road to the river.

While my suggestion was a sarcastic one, it illustrated the absurdity of thinking that tourism would save our small town. It's difficult to capitalize on the visitors who roll past our house in enormous RVs for four months out of the year. The only time they slow down long enough to speak to us is when they need the services of our volunteer fire or ambulance crews.

Tourism has not generated economic prosperity for our area. That's why I'm concerned about another environmental stewardship effort gaining momentum in eastern Montana.

In a movement similar to that of The Big Open project, the American Prairie Foundation has begun establishing a wildlife reserve that it hopes to expand to the scope of the African Serengeti. Because the group has the support of The Nature Conservancy and the World Wildlife Fund, in addition to the support of state and federal government agencies, more than 130,000 acres of land is now controlled by the APF.

I have questions. I wonder how we will keep this area free of brucellosis when the free ranging bison population is fully established. I wonder how young farmers and ranchers will be able to compete with the millionaires who fund these conservation groups when a piece of land comes up for sale. I wonder just when we will see the tourism benefits from these projects.

I wonder why the websites of some of these groups have "human integration" plans in which they discuss training current residents of the area in aboriginal methods of coexisting with nature. I wonder if the people who write for these websites have ever coexisted with nature in a Montana blizzard in February. I wonder how many centuries humans have to inhabit an area before we become a recognized species.

Most of all, I wonder how people are convinced that projects such as these are saving Montana.

To me, Montana is about more than just the wildlife species, the plants, the dinosaur bones, the mountains, and the Breaks. It's more than scenery. It's more than hunting. It's more than farming or ranching or recreating.

To me, Montana is about small communities who are more like extended family than neighbors. Providing African safari-like hunting trips in eastern Montana or blocking off millions of acres so the buffalo can run wild and people can pay to gawk at them is not going to save Montana. As more land is eroded away from private ownership and into wildlife preserves, the number of families living here will dwindle.

Tourists who are here for a few days do not sustain our communities. The people whose blood, sweat, and tears stain the land sustain them. I wonder, then, just who is fighting to save Montana?

The Farmer and the Cowman

Back in high school, I had a brief infatuation with Rodgers and Hammerstein musicals. I suppose this interest stemmed from my love of music, my romantic nature, and the fact that we had three television channels that were usually only tuned into the local news or M*A*S*H.

Among my favorite musicals was "Oklahoma!" It was a natural pick for me because it involved agriculture and featured a strong willed female character who had a tendency to swoon over cowboys, which is a basic description of myself during adolescence.

The highlight of the production for me is the song "The Farmer and the Cowman Can Be Friends," which introduced a concept quite familiar to those who have studied Western culture. The cattle herds in the West were built long before the fences were, and the establishment of farms along with the invention of barbed wire caused a culture war that has never completely subsided. Even today, the term "farmer" and "cowboy" conjure up quite different images.

Cowboys seem to have the advantage in popular culture, where they are admired as independent, hard working, handsome figures who are always polite to the ladies. Farmers, on the other hand, seem to struggle with a few image issues among the general public.

Farmers tend to be viewed as slightly bumbling, uneducated clods who are clothed primarily by seed and chemical companies. They sport pliers in their holster instead of a gun. How many little boys want to dress up like a farmer for Halloween? Unlike the romantic cowboy whose silhouette graces the sunset as he rides off on his steed, the farmer chugs away in his tractor and is blamed for excess emissions that deteriorate the ozone layer.

The farmer earns the wrath of the American public because he accepts government subsidies, applies chemicals to his crops, and slows down traffic while he moves equipment from field to field. But it's hard to hate a cowboy, a character so indelibly preserved in our country's lore.

Despite their obvious differences, farmers and cowboys have been moving toward more common ground. The occasional grumble of cows trampling a crop is still common out here where cowboys and farmers are neighbors, but the fact that farmers and cowboys are both making a living from the land in a world that no longer understand much about production agriculture bands them together. With common concerns of market volatility, weather volatility, and input cost volatility, their similarities probably outnumber their differences.

In fact, in many cases, the farmer and the cowman are not only friends, they're the same person. The rosters of the state's grain organizations and stockman's organizations have many duplicate names.

In my experience, this diversification may make sense from a financial standpoint, but it also creates an entirely new set of conflicts. For example, should he choose baling hay or harvesting wheat? Calving or seeding? Gathering cows in the fall or putting in winter wheat?

At least in this scenario, there won't be any gunfights over the cows getting into the crop.

Shane and Erin Slivka, 2011

Wedded Bliss

June is always the season for weddings in central Montana, and this year is no exception. As the invitations started showing up last month, I started thinking about how many of these young couples would be starting their new lives in agriculture. I couldn't help but wonder if those new brides knew what they were getting into.

Their first mistake was planning a June wedding. It's a traditional month for weddings, and the weather is usually beautiful, but I don't think these June brides of ranchers realize their true honeymoon destination.

It's on the swather.

If you're marrying an accountant, a spring wedding is perfect. You will have the whole summer together to lounge in the yard and barbecue together.

If you're marrying a rancher, you will spend your summer on various pieces of machinery, running lunches to the field, and wondering how long the casserole will stay moist in the oven while you wait for your husband to come in at night.

The closest thing to romance you will probably experience during your first months of married life will be riding on the tractor next to your new spouse, trying to carry on a conversation as you're wedged between the arm rest and the window.

Some of the most meaningful conversations of my marriage have been conducted in such a fashion. We discussed my career path as my head rhythmically bounced off the cab of the tractor. We discussed future children one evening while summer fallowing, but I was distracted by the throbbing pain in my tailbone where the armrest had inflicted its wound.

Of course, it's also awfully romantic when you work with your new husband, too. I feel sorry for those new brides who didn't grow up in agriculture. At least I was used to the long hours, the burnt casseroles and the lack of the fairytale honeymoon.

I also knew the secret language of ranchers that many new wives must painstakingly learn over the first few years of marriage. You see, as a rancher's spouse, you are automatically required to read your spouse's mind. You also must know the sign language that is necessary for those who are moving livestock.

When your mate declares that it is time to move cows and leaves in a vehicle, you must already know several things. First, you are required to know where the cows currently reside and to which pasture they will make their new home. You obtain this information by reading your spouse's mind.

Second, you are required to know the route by which your spouse will move the cattle. This route varies each time the cattle are moved, so again you must read your husband's mind to obtain the necessary information.

Third, you are to know which gates to open and which to close. This is where mind reading is absolutely critical. New wives always open and close the wrong gates. That's a given. Be reassured that, when the situation arises, there is a backup plan.

Your husband, who is 3/4 of a mile away at this point, will

begin the secret sign language that only ranchers understand. You are to recognize the signals and act appropriately according to your instructions. You must never, never think that it would have been easier had he just told you what to do before leaving the kitchen table. That is not an option.

After many years, you will learn to decipher the sign language and to read your spouse's mind on a regular basis. At that point, the cursing involved with moving livestock will be greatly diminished.

Another requirement of new wives is that they know how to operate the ranch's equipment. Husbands rarely have the time or patience to teach their wives how to operate the equipment, and even under the best instruction, a new wife is going to make terrible mistakes.

Case in point: As a rancher's daughter, I never operated a tractor until my first summer as a farmer's wife. When the time came for me to learn, my husband was at my side, teaching me the fine art of plowing. Just when I thought I had it mastered, the inevitable occurred. Never mind the details. Suffice it to say that a new road now existed through the best winter wheat crop on the place.

A new wife always gets put to work during harvest, usually driving the oldest grain truck on the place. A rancher logically chooses that job for his wife because he knows that she can do little to damage the 1952 truck with the crooked box and the bent fenders.

Meanwhile, he speeds off in the air-conditioned combine, barking orders over the radio as you pick grasshoppers out of your hair and wipe the dirt and sweat off your neck. One word of advice: always check the emergency brake when you take off, and if you smell smoke, stop. You'll never live it down if your husband must haul winter wheat seed in a truck with no emergency brake (not that I know from experience, of course).

After reading this column, many June brides may be shrieking in terror. In order to avoid canceled weddings, I will give you just a few of the reasons why I wouldn't trade this life for any other. Even though the tractor is not most new brides' vision of a romantic getaway, it can be a place for ultimate truth and closeness, which no romantic restaurant can come close to providing.

Starting a new marriage on a ranch is more than just starting a marriage. It's building a life and future with one another and with the land. That bond will last long after the newlywed bliss is faded, and it becomes a commitment that will weather the fiercest storms.

A ranch wife may never get a bouquet of roses delivered from the florist, but a bouquet of wildflowers preserved in a lunch cooler and brought home at night will melt her heart. Watching a spring rain together is a sweet moment of success, and watching a hailstorm together creates an even stronger bond between a couple.

It's a life of pain, worry, frustration, helplessness and hard work. It's also a life that can create a marriage stronger than any other, and it's the only life I want to live.

To Love, Honor, and Cherish

As my husband and I approach our sixteenth wedding anniversary this spring, my mind has drifted back to the day we exchanged our vows. As I recall, it rained the entire day right up until the point when it started to snow. Some brides would be distressed at this turn of events, but as a young couple headed toward a future in agriculture, we took the moisture to be a good omen.

Our wedding was not an elaborate event, but we shared the day with many good friends and family members who did a marvelous job of hiding their concerns about our marriage. Not only were we getting married too young, but we were skipping the honeymoon and heading straight into debt on the farm. We carried on with youthful optimism and recited the traditional vows about loving, honoring, and cherishing.

Now that our youthful optimism has evolved into jaded realism, I have realized that the true tests of our marriage were not included in our wedding vows. Sure, there were the mentions of sickness, health, riches and poverty. But after surviving 16 years of an agricultural marriage, I believe I could write a more realistic set of vows that would read something like this:

I, bride, take you, groom, to be my lawfully wedded husband, and I fully understand that I'll be married to your farm as well. I will have and hold you even when you have been bathing in diesel fuel or doctoring scoured calves.

I will love you in good times, like that one year in 20 when the grain market is up and we dodge the hailstorms as well, and in bad times, like when the seven year drought lasts for nine years. I'll have you for richer, when the calf check is in the bank, and for poorer, the day after the calf check is deposited and it's suddenly gone to make payments on the loans. I will believe you when you explain that true wealth is measured in your debt to asset ratio.

I promise to love you even when you track mud across my kitchen floor three times within an hour. I will honor you even as you yell at me across the corral. I will cherish you when you're cursing the weather forecast. I will love you in flooding and in drought, through grasshoppers and bumper crops, in east winds and Arctic clippers, in sorting pairs and cutting out bulls, and I'll stick with you until you retire at the time of your death.

* * * * *

I, groom, take you, bride, to be my lawfully wedded wife, and I understand that you're going to expect me to come in at a reasonable hour to eat every now and then. I will have and hold you even when you have been artificially inseminating cows or pitchforking placentas out of the barn.

I'll have you for richer and for poorer, and I will try not to panic when you spend a chunk of the already spent grain check on new furniture. I promise to love you even when you burn up the parking brake on a vehicle for the second time in a year and when you bring home the wrong parts . . . again.

I will honor you even when you put health food in my lunchbox. I will cherish you even as you're letting the wrong cow through the gate. I will love you in a messy house or a clean one, through postpartum weepiness and post potty training joy, in insulated coveralls or a nice black dress, and I'll stick with you until one of us is gone.

* * * * *

Although our vows may not have quite covered all the bases, they seem to have served their purpose. As the years go by, we certainly appreciate those words we spoke to each other, and we realize how truly blessed we are to remain committed, both to each other and to our life in agriculture.

Emma Lou and Mitch

Farewell to a Faithful Friend

The day was a beautiful respite from the icy cold days we had been enduring for several months. Sunshine washed over us, and as we worked, the pile of shed clothing grew from a few pairs of gloves to a heap of coats, hats, and sweatshirts. Across the ridge a yellow neck tie fluttered in the breeze where its seven-year-old owner had dropped it in his quest to hoist a large rock from the ground. He was wearing the clip-on tie at the command of his older sister, who told him it was proper attire for a funeral.

It was not a conventional funeral by any means, but it was a service that none in attendance will ever forget. Each person present, ages five through 38, worked together in the burial, and with the placement of each rock on the grave, another memory was evoked. More than a few tears fell, and more than a few smiles were shared as we buried a friend and family member.

Mitch joined our family about a year after we were married. He was a small ball of black and white fur with a tongue that rarely stopped licking. Our lives were never the same.

We were suddenly pitched into the world of parenthood without the traditional 9-month transition period. We were awakened at night by the sounds of our crying baby, and our lives revolved around feeding and bathroom times. But we truly began to experience parenthood when our Border collie mutt started venturing out into the world.

Before long, we were on a first-name basis with our veterinarian. Mitch's first brush with death occurred a couple of weeks after he became our puppy. He somehow managed to maneuver his nose into the closet door and slide it open, after which he feasted on bar bait that was supposed to tempt the mice.

Bawling hysterically, I called every vet in the phone book. After trying every suggested remedy, we were finally able to induce vomiting, after which we felt both relief and – well – disgust.

Only a few weeks later we were on the phone to the vet again. This time, Mitch's squeals had awakened us in the night. I reached the screen door just in time to see Mitch rolling across the yard with another black and white furry creature, but this one didn't have puppy breath.

I rushed Mitch to the vet the next day after a vinegar bath muted the pungent smell. After finding no puncture wounds, the vet released our precious puppy with warnings about signs of rabies.

Our "lucky" dog went on to survive a rattlesnake bite, the deadly Parvo virus, two falls from a moving vehicle, and the toddlerhood torments of four adoring children, whom he babysat and herded in the yard with a serious sense of purpose.

Over the years, his eyes became cloudy, and his hearing was completely gone except when he heard the sound of a child walking across the yard with chicken scraps that just might fall on the ground before they arrived at their destination.

His deafness caused great consternation to his owners. He was constantly in the way of vehicles, and several times he was nearly the victim of a tractor tire. He no longer heard our commands to go to the house when we were working cows, and he was rolled countless times by planting himself squarely in front of the head catch, just waiting for it to release his next challenge.

His life was not all work and no relaxation. He owned his own recliner in the shop, and that's where he retreated the day that Cow Number 536 dealt him his final blow after he wandered too close to her newborn and couldn't quite escape her wrath. We made him comfortable and said our goodbyes, finding it difficult to imagine life without our companion of 15 years.

We bade him farewell on the ridge over the coulee. His life filled ours with precious memories, and our lives will always feel his absence.

The Parts Run

When a piece of equipment breaks down on a farm or ranch, it always breaks at a critical time. Swathers break during haying. Tractors break during seeding. Combines break during harvest. When it is ascertained that the breakdown cannot be cured with duct tape or Balen wire, the farm wife is contacted.

"I need you to go get parts," the farmer or rancher will declare urgently.

The first time my dad said that to me, I was somewhat excited. The parts store held a kind of mystique for me because I had never actually entered it.

When I was a small child, I thought the parts store was "town." My dad would come in the house and tell us we had to go to town, and after the hour-long drive all I remember is sitting in the back seat of our station wagon in front of the parts store.

It seemed like hours that we would wait for my dad to make his purchases. To my mother, it must have seemed like days. She was forced to listen to three bored and cranky kids taking their frustrations out upon one another.

"Mooooooooooom, Jimmy's hitting me," my sister or I would holler.

"Knock it off!" responded the fatigued voice from the front seat.

"Mooooooooom, Colleen's on my side of the car."

"What did I just tell you?" growled my mother.

As time wore on, the hot vinyl seats would stick to the backs of our bare legs, and we would yearn to run around.

"Sit down and be quiet!" my mom would mutter. "Did your dad get lost in there, or what?"

Finally, he would emerge from the parts store, and we would head home.

Now that I would finally be able to enter the store, I wondered what I would find there that held my father's attention for such long periods of time. I pulled the car up to the familiar building and anxiously entered it. Once inside, I couldn't understand how my dad could spend so long in such a boring place.

After that trip, I realized that the parts store is an entirely different place for women than for men. For men, it is a sort of Mecca. From farms and ranches everywhere, they make their pilgrimages to the place that houses the answers to their mechanical problems.

For most women, it is a building of doom. During a woman's first parts run, she is usually unaware of the mistakes she is about to make. Her spouse has told her exactly what she needs to bring home, and she relays this information to the man behind the counter. He punches the information into the computer, looks up and asks, "Is that an "A" model or a "B" model?"

She panics. Her husband didn't tell her there would be any questions. She was supposed to give the information to the parts guy, grab the part, and attempt to break a land speed record on her way home with the merchandise.

The parts guy recognizes the distraught look on the woman's

face and gives his co-worker a knowing smirk.

"Well, is it an '82 baler or an '83?" he asks.

"'82?" she guesses.

He disappears down the long aisle of boxes and returns with a small package. She signs the slip and heads home, feeling for the first time the doom that the parts store instills in a woman who has just bought the wrong part.

The parts guy is smiling when she returns with the package two or three hours later.

"It's an '83, huh?" he laughs.

It is not so funny to her, since she's the one who encountered the wrath of the rancher who is home watching the rain clouds build up over his broken baler.

The next time, she returns to the parts store a much smarter woman. She has written down the necessary information. Feeling more confident than ever, she passes her note to the parts guy.

I am convinced that parts guys go to special training sessions where they learn a secret language to make farm wives feel stupid. Even if the wife has all the significant information written on a carefully organized note, the parts guy will always look up from the computer and ask a question in a language that might as well be Greek.

"Is it the left galutaclophy or the right galutaclophy that is broken?" he asks in his secret code.

"Right," she replies, smiling. Having witnessed the breakdown, she clearly remembers the side of the tractor from which the cursing was coming.

He throws the necessary part on the counter. She signs for it and dashes back to the field, convinced that she has purchased the correct part.

No such luck. She is beginning to realize that a strange power is working against her. In fact, some would consider it a conspiracy. Whatever the truth may be, the farm wife will rarely, if ever, return home with exactly the right part, and her husband will become exasperated with her.

She can try insisting that her husband call ahead and tell the parts guy what he needs. It's a nice tactic, but like all other methods, it has flaws. When she arrives at the store, she may find that they are sold out of the necessary part. They may have it in two different types, or it might come with or without an attachment.

The parts guy does not possess the same sense of urgency as the farm wife, either. He ambles back among the rows of parts, stopping along the way to joke with another parts guy or to answer the phone. Meanwhile, the farm wife is looking at her watch, thinking about how she is going to get the kids fed and run the part to the field at the same time.

Even if she obtains the right part in a timely manner, that strange power that hovers over the farm wife will probably interfere at some point. For example, on her hasty drive home, she may encounter a highway patrol officer who thinks that a broken auger is no excuse to abandon reason and prudence on the road.

It's a wise woman who takes her sense of humor into the season of breakdowns. In fact, it may be the only thing that preserves her sanity when, in his hurry to put the part back on the grain truck, her husband breaks something else, sending the farm wife back to town.

Pomp and Circumstances

Graduation in a small Montana town is the only event that draws a larger crowd than a basketball game or a funeral. The community gathers to gaze with pride upon the product of many years of nurturing. As the students walk across the stage to grasp their diplomas, the boys uncomfortable in their suits and the girls wobbly on their high heels, friends and relatives wipe away tears.

Not all the tears are shed out of happiness for the graduates' accomplishments. There are also tears of sorrow at the thought of a loved one leaving home. Unfortunately, graduations in Montana have become more and more sorrowful in the past few years. Ironically, the same problem plaguing our agricultural markets is breaking apart families and causing our youth to leave the state for good.

Montanans take great pride in producing the finest wheat, barley, and livestock in the world. Then they sell it for prices that in some cases do not even cover the cost of production. Montanans also produce the finest children in the world. Companies throughout the world seek Montana natives to join their workforce because they are honest, hard working, and well educated.

However, Montana is losing the very product that its citizens work so very hard to produce.

Other states and countries are reaping the benefits of Montana's proudest accomplishment: its children. We raise our children like we raise our wheat: hardy, dependable, and well rooted. But just when the crop is ripe, we watch someone else harvest it.

You cannot blame the young people for leaving the state. They have opportunities elsewhere that are much too tempting for them to refuse. They cannot be expected to stay in Montana and take jobs with half the pay that is offered them by an out-of-state company. We certainly cannot expect a young person to become involved in production agriculture when generations-old farms are selling out due to drought and low prices.

The cycle that this problem creates is devastating, especially to a small community. More and more you see a 60-year-old couple struggling to harvest by themselves because their children are on the coast working in office buildings. Eventually, the ranch is sold to the neighbors, and another family farm seemingly evaporates.

Consequently, the school enrollment suffers. As a result of low enrollment, the teachers at the school are expected to teach more subjects for less pay. Then the school has difficulty attracting teachers, especially when all the education graduates are being wooed by states with bigger paychecks to offer.

These circumstances are playing out all over Montana, and the ramifications are felt beyond the decline of the small towns. The families that this trend breaks apart pay the ultimate cost of the ill economy in Montana. The close-knit family that battles its way through the challenges of agricultural life finds itself torn apart when the children leave home and move out of state. For those families, I will shed a tear at graduation this year.

The Magic Kitchen Table

I have a magic kitchen table.

When we bought it five years ago, I thought it was fairly ordinary. It is solid wood with a plain finish. It has few ornamental details. I selected it because it could expand to accommodate a crowd. I was unaware of its amazing powers.

Yesterday, for example, I returned to the kitchen after folding a load of laundry and looked in dismay at my kitchen table. In the 15 minutes since I had last cleared it off, it had accumulated enough items that I could barely see its surface.

You see, my kitchen table has the extraordinary ability to generate stuff. When I clear it off and leave the room, it can produce a mess in just minutes. Right now its contents include a half-eaten bowl of oatmeal, a spoon, the chicken scrap bucket (dirty), a pair of shoes, a plastic pony, three Matchbox cars, a sock, a baby bib (dirty), three books, a telephone, a coffee cup (dirty), and a newspaper. This list would be rather unremarkable but for the fact that I cleared it off only half an hour ago.

After a few years of contemplation, I have come to the conclusion that its magic powers lie in its most obvious characteristic: its flatness. Any flat surface in my house tends to reach out and grab things from out of people's hands.

The kitchen table is more than just a dumping grounds for stuff, though. It is arguably the most important piece of furniture in a country home.

The kitchen table is the gathering place for neighbors who discuss markets and cattle over cups of coffee. It is where we convince the banker to give it one more year here in next year country. The kitchen table is where we sign our life insurance policies and negotiate calf prices with the cattle buyer.

In the evening, the kitchen table is where the kids gather, almost instinctively, to do their homework and converse with Mom while she is making supper. It is where the multiplication tables are learned, science experiments are constructed, bears are counted and sorted by color, and book reports are composed.

On Saturdays, the kitchen table becomes another type of classroom. The lessons of the kitchen are passed from one generation to the next as noodles are dried, bread dough is kneaded, and biscuits are rolled out. This is where my kids decorate the Christmas sugar cookies, putting on too many sprinkles and breaking off the tips of the stars "accidentally" so they can sample their work.

The kitchen table is the most photographed furniture in the house. I have pictures of my kitchen table surrounded by grandparents, great grandparents, aunts, uncles, and cousins. They are singing "Happy Birthday" and watching children blowing out candles on cakes decorated on the kitchen table late at night. They are eating Thanksgiving dinner or tentatively sampling the kids' birthday cake for Jesus on Christmas Day. These photographs of people around the kitchen table chronicle the passage of time, the growth of children, and the aging of our elders.

The kitchen table is where my children learn to give thanks

to their Father for his blessings. It is where they hear their daddy pray for rain. It is the only time that my oldest child will hold the hand of his younger sister as his daddy asks God's blessing on the meal before him.

The kitchen table is where the kids hear their mom bring up touchy subjects because she knows it is hard to argue when your mouth is full of pot roast.

This is the place where we answer questions like, "What is war?" and "Why were those people in the Bible naked?"

So, you see, my kitchen table is magic not just in its ability to accumulate stuff, but in its ability to gather together the people who mean the most to me and for the opportunity to teach, to learn, and to live.

Riley and Skip, the resident babysitter

Children of Summer

The day has arrived. My children have discovered that I am old.

They had their suspicions when they realized that I was struggling with simple tasks like programming the VCR and setting the microwave clock. They would roll their eyes and shake their heads at one another as if to say, "What a geezer."

Now they are beginning to recognize other signs of aging, such as the fact that the waistband of my pants sits at my waist, not somewhere between my hips and my knees. I do not carry a cell phone on my hip at all times. My sunglasses are several years old and do not cover my entire face. I use the word "bad" in reference to something that is not good.

It wasn't until I began talking about my childhood that my kids realized just how ancient I really am, though.

They were astonished that my third grade classroom did not have a computer and that when I was in fifth grade, our computer had no mouse and disks that really were floppy.

They were shocked that we did not have iPods and that those abandoned glass boxes on street corners used to house phones before everyone had their own cell.

While I have been busy chauffeuring them to activity after activity all summer long, I have explained to my kids what my childhood summers were like.

Even today I can sometimes imagine that I hear the sound of hundreds of sheep mothering up at dusk, a sound that was both raucous and comforting. It is a sound I memorized during the countless summer nights my brother, sister, and I spent in sleeping bags in our front yard. Our beds were virtually unused in the hottest summer months; we instead opted for the cooler comfort of the outdoors, where we would try to stay awake long enough to see the last falling star. As the night grew colder, the animals would move in, and my parents would frequently find us covered by dogs and cats by early morning, when the meadowlarks awakened us with their trills.

In contrast to my kids' overscheduled summers, those three months during my childhood were a time of isolation from the world. Since my parents were so busy with ranch work, we rarely left the place. My friends from school were forgotten until fall, which strengthened the bonds between my siblings and me. After all, you cannot afford to alienate your only playmates.

There was an occasional road trip to visit family or friends, which I recall by the feel of my sweaty bare legs stuck to the hot maroon vinyl in our station wagon and the sound of my brother's whistling, which was maddeningly horrible.

While my kids are bored after just a couple days of no activities, I can remember creating all kinds of games and amusements to occupy our warm summer days. I walked all afternoon turning over bales with the dogs at my side, waiting eagerly to dive onto the mice that occasionally hid beneath the bales. I scoured the same hayfields searching for gopher holes. Sometimes I would pour poisoned oats next to the holes, but often I would set traps, as the reward was a buck a tail for every success.

I read Zane Grey novels in the grain truck, chewing kernels of wheat into gum and waiting for the combine to come back around. My brother and I raced bicycles and explored abandoned homestead shacks. I halterbroke bum lambs and tamed barn kittens. If I ever uttered the words, "I'm bored," my mom quickly produced a solution to my problem, usually in the form of weeding the garden or hand trimming the grass around the house.

Explaining all this to my kids produces quizzical looks and wide-eyed wonder at a time that seems foreign to them. I recognize the look of horror in their eyes; it is the same look I had when my mom explained that not all the houses she lived in as a child had running water and electricity. It is the same look I had when she told me she was required to wear a dress to school every day. It is the look that says, "Wow. You're old."

My fond memories of childhood summers are translated to my children as the tales of a disadvantaged youth who grew up in the dark ages.

Someday, maybe they'll appreciate the memories as much as I do.

Country Kids in the City

$\mathcal{O}$f all the decisions I have made throughout the years, I like to think that one of the wisest was choosing to rear my children in the country.

There are also many times that I doubt my wisdom.

One of the primary benefits I can usually cite for living rurally is that I can protect my kids from some of the harsher elements of urban life. It is that very protection, however, that can sometimes cause trouble when I take my sheltered children to town.

For instance, my two-year-old son does not know that he cannot run out into the street. He doesn't even know what a street is. If he sees a tractor coming across the field, he knows that he needs to avoid it. But since we live well off the county road and rarely spend time anywhere with pavement, he is completely unaware that a street poses any danger to him.

Ignorance of traffic can be a benefit of sorts as well. I always try to obtain a window seat in busy restaurants because my sheltered children are fascinated with traffic. All the city kids in the place completely ignore the cars whizzing by their windows, but my kids sit quietly and stare out the window at the unfamiliar sight.

They are not always so well behaved in the city, though. When my oldest child was three, I was browsing through a clothing store when he suddenly disappeared. I was nearly frantic with worry until I saw a fellow customer jump back in alarm, stifling a scream. There, with his head poking out from a circular rack of dresses, was my son, playing an impromptu game of hide and seek.

I have also found that sheltering the kids in the country may not provide them with all the social skills they need as they encounter people in town. This point was proven when my child pointed at the checkout clerk and said, "Do you have a baby in your tummy, or are you just fat?"

At least that episode provided a teachable moment. Some skills are more difficult to teach, such as the concept that not everyone in a city is interested in who you are or what you are doing. My kids are used to a small-town grocery store in which everyone has known them since birth. They can happily rattle on to the clerk and the customers about their daily lives and accomplishments, and they always find a receptive audience.

Thus, when we are in a large department store, I find myself dragging my daughter away from strangers that she is trying to engage in conversation. She has no understanding of why they would not be interested in her full name, age, interests, and opinions.

Managing naïve and inquisitive country kids in the city can be an exhausting experience, and it is usually at that peak of exhaustion that I begin to question my wisdom in a) rearing the children in the country, and b) bringing such countrified children to the city to begin with. Such was the case last summer when my four-year-old daughter and I were waiting in the parking lot for the rest of the family to pick us up.

My daughter informed me that she had to go to the bathroom, and I was not about to re-enter the busy store that I had finally escaped with a cart full of items. I told her that she would have to wait a few minutes, and I continued to scan the horizon for her father.

When I turned to make sure she was still there with me, I saw the sight that no mother wants to see. There, squatting by a pencil-thin tree in full view of the world, was my half-naked daughter.

She was surprised at my gasp of horror. After all, her brother taught her that when you are outside and you have to go, you simply find a tree and do your business. It's a lesson that his father passed down to him, and it's one that is completely acceptable on the prairies of Montana. In the largest city in the state, however, I believe it is generally frowned upon.

Although I attempted to explain that logic to my daughter, I'm afraid that the lesson was lost upon her happy little spirit. At least the big city provides one amenity not found in rural areas: anonymity.

Anna carries on the family tradition of feeling more at home in the lambing barn than in the house.

A Woman's Place is in the Home

We recently took advantage of one of the five days of summer we've had this year to vaccinate calves. With a good crew on hand to help, the work went as well as can be expected when cattle are involved. In fact, nary a cuss word was shouted throughout the morning, and after a summer of grazing the cattle in the breaks pasture, we considered ourselves fortunate to find only one calf missing.

As I put a late lunch on the table, I was looking forward to a more relaxing afternoon of updating cattle records and catching up on a few household tasks.

No sooner had the thought crossed my mind than my husband announced an alternative plan. He had decided that since the horses were still saddled, it would be a great time to situate the bulls in their fall pasture.

The crew around the table quickly vanished at the mention of moving bulls, and when the dust settled, only my husband and I remained. I desperately looked around for someone, anyone, else to share the burden of the task at hand. I've found that if no one else is helping, it's hard to avoid the blame when something goes wrong.

My mind drifted back to the last time we had dealt with the bulls. The entire herd had stampeded into the neighbor's CRP, and we spent 20 minutes gathering them back up while playing dodgeball with the round bales they were pushing. About 15 minutes after that, they found the hole in the fence where someone had lost a pickup off the road, and later we had the pleasure of laboriously cutting out the bulls only to watch one vault over a closed gate to be reunited with his girls.

Needless to say, I was not optimistic about the afternoon's task. I knew that the pasture to which we were headed was home to a row of grain bins and a sizeable swamp. Our task was to cut out one bull to leave behind and push the others to the gate, across the county road, through the neighbor's yard, and into the open gate.

The open gate, I might add, was leading to the pasture in which we had just deposited the fence-vaulting bull in the midst of his cohorts, who were now snorting and head butting and throwing dirt like. . . well, like bulls. They also happened to be heading straight for the open gate.

Meanwhile, on the other side of the road, things were not going well. It became obvious that we were not going to cut out one bull from the others, especially after the whole works dove into a bog and were threatening to go through a fence. We would have been better off abandoning the horses and resorting to a rowboat; at least then we could use the oars as cattle prods.

The boss decided to take only half the bulls across the road, where by this time the bulls had found the gate and were dancing around the neighbor's yard in a magnificent display of excess testosterone.

We pushed the whole swirling works toward the gate, but two bulls were engaged in such a headlock that they paid no attention to our attempts to shuffle them inside the fence. As we played the game of keeping the other bulls inside the fence and trying to keep the fighting bulls from straying too far away, I wiped away the sweat that was dripping from my head, thinking longingly of the snow and freezing temperatures we had been complaining about only a week before. That's about the time a bull crashed through the fence and headed up the road.

We cut off the obstinate animal before he strayed too far north, but he dove into the ditch and was just beginning to push through yet another fence when my husband began yelling at me to kick my horse and go in after him. I looked down over the precipice, out of which was jutting a large metal culvert, and decided to defy my orders.

I may have said something like, "Are you nuts?"

The culvert had been doing its job, and the entire ditch was underwater. "We're gonna lose him!" was the response from the boss. I didn't bother to mention that, at this point, I didn't care if the bull ended up in North Dakota. I was not diving into a ditch full of water to face off with a bull full of snort.

We managed to regain control of the bull and push him back to the gate, and I noticed that my husband's tone improved significantly after I had the opportunity to chase down his loose horse, catch him, and hand over the reins. While our mission was not completely accomplished, at least most of the bulls were where they belonged, and if nothing else, we were still married.

As I reflected on the day's events, however, I announced to my husband that I have had a revelation.

Next time he needs help with bulls, I'll be quick to inform him that a woman's place is in the home.

Before You Take Another Bite

Our family took a rare day off recently and headed to town, stopping to eat at a kid-friendly fast food restaurant. Although we tucked our boisterous bunch into a corner of the room, we were still situated rather close to another table where a couple were reading the paper and discussing the weather and NASCAR. I smiled, thinking how fortunate we are to live in an area where people are so down to earth.

My smile quickly faded, however, when the subject suddenly changed and the man began to loudly rant about farmers, ranchers, and wolves.

Between bites of his hamburger, he was animatedly describing how the land owned by farmers and ranchers should really be turned back to nature.

"Farmers aren't making any money anyway," he bellowed. "We ought to just take that land away from them and preserve it for the animals. All the ranchers want to do is just to kill all the wolves."

My blood pressure began to rise with each word he uttered, and I caught my husband's eye just as I was about to come uncorked. He silently told me to stay in my seat, and I knew he was right. Engaging this individual in an argument would prove futile.

I longed to explain the concept of irony to my children as we sat listening to a man with a belly full of hamburger berate the people who made it possible for him to stuff his face. I wanted to ask him if he was enjoying his meal and see his reaction to my question. I really wanted to ask him where he would be eating if all the land were turned back to wildlife and our agriculture-based economy were to disintegrate.

Instead, I used the experience as an opportunity to utilize the Lamaze breathing techniques I learned many years ago. Patience is not my strongest virtue, but I did an admirable job of biting my tongue and waiting for him to stop berating our occupation. Once we were safely ensconced in our vehicle, I turned to my husband and expressed my frustrations.

I decided it might be helpful to people like my fast food restaurant neighbor if I were to clear up a few misconceptions in a simple, concise manner.

That hamburger you are eating is made from beef. We raise beef on our land. Wolves like beef, too. If we can't control the wolf population, there will be much less beef, and it will be more expensive. How do you feel about tofu?

The bun on your burger is made from wheat. We grow wheat on our land. If you follow through with your brilliant plan to take the land away and give it back to "nature," no one will be there harvesting wheat that can be made into bread.

The fries you just swallowed are made from potatoes that are grown by. . . you guessed it, farmers. The same farmers that you would like to put out of business. How does your plan sound now?

That drink you just slurped is full of sugar grown by farmers.

The lettuce and tomato on your burger were grown by farmers.

Do you see a pattern here?

And by the way, would you please share a few things with your anti-agriculture cohorts who want to turn millions of acres over to the federal government and create a free ranging bison preserve?

There are, in fact, people who live in north central and eastern Montana. These people do, in fact, own land within the Missouri River Breaks National Monument and in the area targeted for a new national monument. These people would, in fact, very much like to stay there and continue making a living in agriculture. So please dispel the rumor that only public land is affected by monument designations. It's not true.

Turning bison loose in the middle of cattle ranching country is a bad idea on many levels. Yes, I realize that bison once roamed free in Montana. But the bison, like outhouses and the Pony Express, would be a poor fit for our modern way of life. The clock can't go backwards, and it shouldn't.

We understand why you feel so strongly about preserving our land. Many of the open landscapes in this country are clogged up by development. It's nice out here with the fresh air and the room to breathe it. But before you lobby to turn it all over to the government to manage it, maybe you ought to consider that someone is managing it right now. And if you'd ever like to come for a visit and find out how we keep it this way, you're welcome at my kitchen table any time.

Just remember where the food comes from before you take another bite.

St. Wenceslaus Church, Danvers, Montana

Disappearance

The boy was restless. They had been driving for what seemed like an eternity, and before that, they had spent the day in two separate airports. When they landed in Billings, the boy had been relieved to finally reach Montana, but he hadn't realized how far from their final destination they still were.

His father's mood had grown more somber as each hour had passed. He would occasionally slow the rental car and look intently out the window, as if he were searching for something.

The boy looked out the window as well, but all he had seen in the past 85 miles was a sea of sagebrush and an occasional sign reminding travelers that this was federal land and that offroad travel was not permitted.

After what seemed like an eternity, they reached Malta. The weary travelers exited the car and stretched their legs in the parking lot of what appeared to be a tourist attraction with tipis circling around them.

The father walked into the largest of the structures while the boy unloaded their luggage from the back of the car. He tried to look past the teepees and into the town beyond, but darkness had fallen, and all he could see were streetlights and quiet sidewalks.

His father returned with a key, and together they carried their bags into the tipi to their left, which a dangling brass number nailed to the side of the door identified as number 7.

The boy rolled his eyes and asked his father if they really had to stay in such a hokey tourist trap. A shadow of sadness crossed his father's eyes as he turned to his son.

"It's the only place in town," he replied.

The next morning, the man at the front desk in the largest tipi directed them to a tour guide who was loading a case of bottled water in the back of an SUV. They were joined by a couple from Connecticut who were exclaiming about the wide open spaces as drove north out of town. The tour guide was quick to explain that this entire area had been rescued by the federal government when the President used the Antiquities Act to designate it a National Monument.

"Together with the grassland reserve in Canada, this area encompasses the world's largest wildlife refuge," the guide explained. "Bison were released here in 2019, and since then, their population has grown to nearly the level it was prior to the settlement of the West."

As if on cue, the group spotted a bull rubbing his head vigorously on the remnants of a home long since vacated.

They continued their journey northeast, and the boy was disappointed to see the same sea of sagebrush they had seen the day before. He began to question why his father thought it was so important that he visit this area. And he began to wonder why his father had such good memories of growing up here.

After consulting the coordinates in his GPS unit, the tour guide stopped the vehicle and told the group that they would be walking the rest of the way to the ranch since federal law prohibits the use of motor vehicles beyond this point.

Thirty minutes later, the guide pointed to a dilapidated house and a hip roof barn. He told the group that the ranch was once the home of five generations of the same family who eked out a living raising cattle. They decided the life was too difficult in 2016, however, and they sold the place to the government. Since that time, the guide explained, the ranch had been turned back to nature, and under federal management, it was preserved and managed for the benefit of the wildlife who resumed their rightful place on the land.

At that point, the boy's father looked piercingly at the guide, and to the boy's surprise, he began to speak for the first time that day.

The boy listened intently as his father told the guide that when he was a boy, his father had taught him about conservation by example, not by government guidelines. He related how his dad had guided the swather around a patch of alfalfa because he knew there was a nest of pheasants there. He remembered walking past the haystack on his way to the school bus stop and seeing the deer bedded down between the big rolls of hay that sustained them during the winter. He talked about the venison they ate for dinner that night as his dad explained that if they provided for the wildlife, the wildlife would provide for them.

The boy's father went on to explain how they had changed farming methods through the years, relying less and less on tillage and chemicals as technology progressed and allowed them to keep the land healthy. He told them about the workshops his father attended to improve the genetics of his cattle herd and make the cows more efficient. He spoke of how noxious weeds were eradicated using biological control methods along with chemical control until they were no longer a problem.

He spoke of how his grandmother had sold eggs and milk to

buy enough paint to finally cover the whole house that was falling down before them. The boy listened closely as his father explained what really happened when the government bought out his dad's place. His family, he explained, had been leasing a portion of their grazing land from the government for generations. When nearby land was declared a National Monument, the government had offered a large sum to buy out his dad's ranch. His dad, committed to ensuring the survival of the local school and community, had flatly refused.

Two years later, the family's grazing rights on the public land were revoked. In the years that followed, the boy's father explained, the family sold half their cattle, and his mother went to work in town, driving 40 miles each way to earn enough money to put groceries on the table. But the same year the grazing rights were revoked, an animal rights organization successfully lobbied Congress to pass new regulations regarding the beef industry, and feedlots all over the United States were shut down. With no one to buy their calves, the family had little choice. The place was sold to the government, and the family moved to town.

The group fell silent as the boy's father looked at his son.

"This is the land I wanted to pass along to you," he said. "For 130 years, our family called this land home. We belonged to this land just as much as the eagles or the deer or the sage grouse. But somewhere along the way, people became a forgotten part of the ecosystem. Government decided it could conserve this land better than we could. Our family disappeared from this land. Our community was eradicated. Our state became a retreat for the wealthy. Our culture was lost.

"But I wanted you to see, son, what you could have called home."

Flax Schoolhouse, Winifred, Montana

Sainer cabin, Winifred, Montana

Barn loft at the Robertson Ranch, Judith Gap, Montana

Colleen, Erin, and Jim Robertson
Judith Gap, Montana

Made in the USA
Lexington, KY
19 July 2015